COMMUNICATIONS HANDBOOK

Order from

THE INTERSTATE
Printers & Publishers, Inc.

DANVILLE, ILLINOIS 61832-0594

COMMUNICATIONS HANDBOOK

Agricultural Communicators in Education

FOURTH EDITION

COMMUNICATIONS HANDBOOK

◀ *Fourth Edition* ▶

Library of Congress Catalog Card No. 81-85216

ISBN 0-8134-2226-4

Preface

"Many of the conflicts of our times will be resolved when we learn better how to communicate with each other. The **COMMUNICATIONS HANDBOOK** is dedicated to that purpose. It is for those men and women who are devoting their professional careers to the important task of communicating educational information to people, especially in the broad fields of agriculture and home economics."

This statement, appearing in the first edition of the **COMMUNICATIONS HANDBOOK**, is as true today as it was when it was written.

The chapters were compiled by agricultural communication specialists. Their orientation to communications and their use of communications media has been primarily in the subject matter of agriculture, home economics, and youth work. However, the instructions given in these chapters are suited to all agencies and activities where the concern is to communicate useful information or educational material to a large number of people.

The handbook itself was modeled after a communications handbook published by the Office of Agricultural Communications of the University of Illinois College of Agriculture.

Members of the Illinois staff supervised the editorial production of the original AAACE handbook. The individuals responsible for directing the editorial production of the second edition were John W. Spaven and his colleagues at the University of Vermont. Eldon E. Fredericks headed the editorial production of the third edition.

The organization of the fourth edition of the **COMMUNICATIONS HAND-BOOK** is essentially the same as previous editions, with the addition of Chapter 13, which recognizes the rapid development of new communications technology, and Chapter 14, which stresses that the most effective communications come from a coordinated approach to communications using all appropriate channels. In all chapters, the content has been revised to reflect the latest information on principles and practices of using specific communication techniques.

Many people have contributed to the revision of this fourth edition. In some instances, technical media committees of Agricultural Communicators in Education (ACE), formerly the American Association of Agricultural College Editors (AAACE), which is the professional society of agricultural and home economics communication specialists associated with the 68 land-grant universities, the U.S. Department of Agriculture, and commercial media associations, have participated. However, credit has been given to the person(s) with the primary responsibility for writing or revising the respective chapters and to the contributors of illustrations for each chapter.

Also to be acknowledged are Cecil D. Nelson, Jr., University of Minnesota, who designed the cover; Bill Ballard, North Carolina State University, who provided a number of drawings and assisted with design and layout; and Patricia A. Ward, The Interstate Printers & Publishers, Inc., who edited the fourth edition and designed the handbook's new format.

William L. Carpenter
North Carolina State University

Contents

1 COMMUNICATION CONCEPTS

◀ The Communication Process ▶

Let's say your day began this morning with the ringing of an alarm clock. As you reached over to shut it off, you reacted in several different ways.

You looked toward the window and saw rain hitting the pane. Immediately you began making adjustments for the remainder of the day.

As your feet touched the cold floor, you reached for your slippers beside the bed.

As you finished dressing, you caught the welcome odor of bacon frying and mentally checked the number of minutes before breakfast would be on the table.

Your day is only a few minutes old, but already you've been intimately involved in a whole series of communication experiences.

Yet, not a word has been spoken.

You've used four of the senses—hearing, when the alarm awoke you; sight, when you saw the rain; touch, when your feet reached the cold floor; smell, when you realized bacon was cooking. You'll use your fifth sense—taste—when you eat the bacon.

In each case, some source initiated a message and sent it through some channel to you so that you could receive it. Such a series of events is known as the process of communication. Fig. 1-1 diagrams one graphic view of the process.

We all know that communicating is not easy. Even though our diagram is simple, each communication situation can become complex and difficult.

We tell a friend to meet us "*on* the corner *by* the drugstore," only to find our friend (after waiting half an hour) "*in* the corner *of* the drugstore."

To communicate successfully, we need to be familiar with the communication process and all the factors involved. Knowing these factors can help us plan, analyze situations, solve problems, and in general, do better jobs. Some factors give us specific ideas to think about when we try to communicate with different people.

Important to remember is that the process is in operation all the time and certain segments of the process may represent longer periods of time than others. The process includes a **SOURCE**, a **MESSAGE**, a **CHANNEL**, and a **RECEIVER**.

Source can be anyone, any group, or even any institution that can initiate a message. Several things determine how a source will operate in the communication process. They include the source's *communication skills*—abilities to think, to write, to draw, to speak. They also include *attitudes* toward the audience, toward the subject being addressed, toward one's self, and toward any other factor pertinent to the situation.

Knowledge of the subject, the audience, the situation, and other background also influences the way the source operates. So will social background, edu-

Prepared by **Hal R. Taylor,** former director of public affairs, U.S. Department of Agriculture, with assistance from Mason E. Miller, communication scientist, SEA–Cooperative State Research Service, U.S. Department of Agriculture; Donald F. Schwartz, chairman, Communication Arts Department, Cornell University; K. Robert Kern, former extension editor, Iowa State University; and Paul Gwin, former information specialist, University of Missouri.

cation, friends, salary, culture—all sometimes called the *socio-cultural context* in which the source lives.

Message has to do with the package to be sent by the source. The source should consider several sub-factors.

The *code*, or language, must be chosen. Generally we think of code in terms of the "accepted" languages—English, Spanish, German, Chinese, and so on. Sometimes though we must use other languages—music, art, gestures, and so on. In all cases, we need to look at the code in terms of ease or difficulty for audience understanding.

Within the message, we need to select our content and organize it to meet acceptable *treatment* for a given audience or a specific channel. If the source makes a poor choice, the message likely will fail.

how well a receiver can hear, read, or use other senses. *Attitudes* relate to how a receiver thinks of the source, of himself/herself, of the message, and so on. *Knowledge* may be more or less than the source's knowledge. *Socio-cultural context* may be different in many ways from that of the source, but it will be made up of the same factors. Each will affect the receiver's understanding of the message.

So when you stop to think about it, there are many reasons why messages sometimes fail to accomplish their purpose. Frequently the source is unaware of receivers and how they view things. Certain channels may not be as effective under certain circumstances. Treatment of a message may not fit a specific channel. Or, some receivers simply may not be aware of, interested in, or capable of using certain available messages.

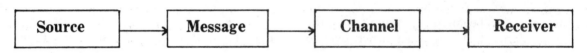

Fig. 1-1. A simple model showing the basic steps in communication.

Channel can be thought of as the senses. Sometimes it's preferable to think of channel as the method over which the message will be transmitted. In telegraph transmission, channel would be the wire over which the message is sent. It also may be the newspaper in which your column appears or the radio station and air waves over which messages are carried.

Kind and number of channels to use may depend largely on purpose, but generally the more channels we can use, the more effective our message. For example, movies or television may be more effective than a book if our purpose is entertainment. But when senses are stimulated directly, say when a first sergeant yells at a recruit, the message is reinforced. Direct stimulation sometimes involves face-to-face contact or an actual experience for a receiver. Several different senses are stimulated.

Receiver becomes the final link in the communication process. The receiver is the person or persons who make up the audience for your message. All the factors that determine how a source will operate apply to the receiver. You may want to refer to them. *Communication skills* might be thought of as

Perhaps that's why some of us failed to react when our alarm rang this morning. And when we told that friend to meet us "*on* the corner *by* the drugstore," only to find that the message became "*in* the corner *of* the drugstore," we may have found ourselves wondering about two aspects of communication: communication accuracy and communication success.

What we do to get accuracy is one thing. What influences success may be another—or at least many other factors may be involved.

Let's back up a bit and think more about the communication process.

Most communication experts agree on four points about the process. First, not only is human communication a process, but it's also an on-going, dynamic one. Second, the process is irreversible. Original impressions can't be recalled or erased. Third, someone's perception is necessary to communication. Fourth, communication takes place within a situational context.

Beyond these four points, the experts disagree or have slightly different points of view. For instance, they ask: How many people must be involved? Are we restricted only to messages intentionally sent forth by a source? Do we include both verbal and non-verbal messages? And if non-verbal messages are allowed, are they restricted only to those that accompany verbal messages?

One point of view holds that communication takes place when only one person assigns significance or meaning to an internal or external stimulus—such as our morning alarm clock. Other experts say two people—a source and a receiver—are necessary for human communication. Still others argue that regardless of the number of people involved, communication can't be different from all other kinds of human behavior unless there is some intention by a source.

Should we include non-verbal messages? They relate to the issue of intention, since much unintended information transfer comes about by non-verbal methods.

Suppose your child throws a pad of butter against the wall. There's a message for you in that. It also may have been intended. Your child and any other children around who saw what happened expect you to react, maybe to do something. What you do or don't do, whether you react as they expect, has meaning. And that happens even though you may have intended to create no message at all. In this situation, the air is electric with all sorts of messages, even though no one has said a word.

If we insist that silence or inaction means nothing, we've denied the idea that communication comes about only through symbols—anything that stands for or represents something else. But if we can accept the idea, we can see that both verbal and non-verbal behavior are involved, that at least two participants are necessary and that communication consists of both intentional and unintentional messages *and* unintended interpretations.

If we accept these ideas, then we can go even further and say human communication takes place when one person perceives and assigns meaning to the behavior or behavioral products or records of another person.

The value of this point of view is that it emphasizes the importance of the receiver in the communication process. We are forced to consider communication from the receiver's point of view rather than emphasizing our own skills and strategies from our source point of view. We are made aware that what we—as a source—intend to communicate may not be received at all in the way we want it to be received. We can understand that meaning may be different for different people. We can become more accurate when we communicate. That is, we can be

better at getting the meaning in our heads into the heads of receivers with a minimum of distortion.

We can't say communication accuracy is the same as success, because, odd as it may seem, sometimes we may not need or even want accuracy to obtain success. There are times when it may be in our best interest as a source to be deliberately vague, if only to gain feedback from others or to "test the waters," as politicians often put it.

Communication accuracy is source-oriented. It is a goal we as message senders give much attention to. So is success—having the effect on the receiver we want. But to assess success in communication fully, we need to focus on effects and purposes that aren't necessarily only our own.

Maybe we merely want to create awareness. Or, we may want to increase the level of information to some audience. On the other hand, perhaps we want to change attitudes, to move others from a neutral position in favor or against. Or, we want to reverse an attitude—pro to con or con to pro. It is possible we simply want to modify an attitude either by strengthening a position or by weakening one. Or, perhaps we want to change behavior or to reinforce it.

We might want to do one or a number of these things. No matter, receivers may think otherwise. It's a certainty that there most likely will be more than one effect, and many of them may be unintended. Needless to say, that's why it becomes so complex to assess whether or not a communication effort has been a success from our viewpoint. Sometimes the only sure way to find out means we must establish formal studies to see what the effects have been.

If we focus on effects and purposes that are not just our own, and agree that human behavior is purposeful and motivated, then we can also see that receivers have intentions or purposes when they direct their attention to a given message. That gives us a second and equally important basis for determining communication success—success from the receiver's viewpoint. Again, that's available by formally comparing the receiver's purpose for paying attention to our message to the obtained effect in the receiver. That further complicates assessing success in a communication situation, but it facilitates thinking about communication planning.

"Well," you say, "now all I have to do is set up some research projects." Not necessarily.

Emphasis on receiver behaviors, not just what *we* have done as sources, can be handled fairly well by observation. The old phrase, "Know your audience," begins to make more sense.

If the secret of communication is knowing people, then the unsuccessful communicator probably is one who doesn't know his/her audience. He/she may be able to write, speak, and take pictures skillfully, but he/she doesn't know or hasn't taken the trouble to find out "what makes people tick."

WHAT DOES MAKE PEOPLE TICK?

To answer the question, we need to remember that human beings are constantly engaged in making adjustments to the presence and activities of other human beings. We must also recognize individual drives, motivations, and frustrations.

Since we can think, we can also deal with abstractions. These give us certain desires, wants, or wishes. And these we must satisfy.

In other words, we have certain motivations which move us to act. These fall into four groups: (1) security, (2) response, (3) recognition, and (4) new experiences, Although these are simple enough terms, not everyone responds to them in the same way. And, it may seem too pat to limit motivations to these four categories. After all, surely some motivations have altruistic beginnings—to do good, to contribute to society, to help humankind. So look at the four groups listed here to get an idea of what motivation is all about and how it can be accounted for in communication planning.

THE WISH FOR SECURITY

Our wish for security has many interpretations. Adequate food, clothing, and shelter generally come to mind first, because nearly everyone wants protection and comfort for his/her physical being. But money and wealth mean security to some, just as do other items adding to material well-being. Some people take satisfaction in thoughts of a spiritual hereafter. Security also means security within a group. And often, when our drive for security is strong enough, we accumulate great wealth or large amounts of material possessions.

THE WISH FOR RESPONSE

Unless we're really different, we want others to like us. We want others to extend themselves to us, to need us, to appreciate us. Such a need is called the wish for response.

As a motivation, the need for response leads us to do what others expect us to do—what our parents, friends, neighbors, children, and associates want us to do.

When we succeed in our drive for response, we know that we are "part of the crowd." Young people especially have a strong need for response. They would rather be scolded than ignored. Few if any of us ever outgrow this need completely.

The desire for response differs from the desire for recognition. With response, we don't particularly care whether or not we are richer or poorer or ranked in a certain position. We just want to be liked and appreciated.

THE WISH FOR RECOGNITION

Some people want to be different. They want the highest crop yield in the community, the best hogs, the fattest cattle, or the cleanest yard. Frequently when we have these desires, we are seeking security. More often, we are motivated by the desire for recognition.

The person who wants recognition expresses that desire through entertainment, tradition, keeping up with the Joneses, status demands, competition, and so on. We want to be somebody and to move to the top.

Often our wish for recognition opposes our wish for response. Being at the top of the heap may not mean that people will like us.

THE WISH FOR NEW EXPERIENCES

Many people occasionally grow bored with old routines. They want to try something new and different. Some have this desire more than others.

Some apparently don't want recognition. Security may mean little or nothing to them. They may, to some degree, want response. But, most of all, they want the thrill of something new, something different.

When we satisfy this wish, often our senses play a large part in our drives. Maybe we like to travel. Or, we can't live with what we have in our homes and are forever changing furnishings. Perhaps we like to try new foods, begin new hobbies, seek out or create new social situations. We seek contacts with new people, look for and accept new and different responsibilities.

Success in getting others to "try" a new practice may depend on the strength of their desire for new experiences.

DRIVES DETERMINE DIFFERENCES

Everybody has these basic drives, or wishes, in different degrees and combinations. Sometimes the forces themselves are in conflict, and as we get older or as our situations change, the relationships of the drives also may change.

Thus, we are a society of individuals, each different from the other. The degree of difference is based on the importance each of us gives to the four motivating forces and to the various interests we have which make us act as we do.

Because we as humans need other humans to exist, we also find that we can best attain and fulfill our basic drives through group activity. When individuals give reasons for joining and participating in groups, they are expressing their basic drives in terms of specific interests.

USING MOTIVATIONS IN COMMUNICATION

As a communicator, you have the task of deter-

Fig. 1-2. Human characteristics, and individual drives, motivations, and frustrations among individuals, must be taken into account when we plan our communication activities.

mining which of the motivating forces are strongest among the people of your audience. Think of them as "appeals" to use when you write a news story, prepare a radio or television program, or hold a meeting. It is not enough just to present information. Put it in terms of one or more of the basic needs or desires of the individuals in the audience, and your chance for success will be better.

◀ Individuals in a Group ▶

In our work and in our leisure we spend much of our time with groups of people. That's because we generally need others to fulfill our own ambitions and aspirations.

At the same time, groups help set our standards. They tell us how far we can go as individuals—at least as members of a specific group.

We belong to some groups automatically. Included would be our family and the organization for which we work.

But for the most part, we belong to groups because we want to. Doing so often satisfies one or more of our individual motivating forces—security, response, recognition, new experiences. Belonging to a group often becomes a very necessary way to reach some or all our goals.

No matter why we become group members, we still become a part of that group's structure. Within it, we'll have a particular role, a certain status, and specific rights and authority. We accept responsibility. The values that the group holds to be important

will determine the structure that it has, and consequently the relationships that prevail within the group or system.

If that is true, then why do we have conflict within groups?

As with individuals, a group or social system often will have values which are in conflict. Goals may appear contradictory. As members of the system learn more about the group, they make judgments. These become a part of the values to which they will cling until they make new judgments and have different values.

Sometimes we rank each group in terms of our agreement with its values. We may not express this ranking explicitly, but we do it.

Then all of us usually orient our own personal actions in terms of the systems or groups to which we belong and in terms of those we consider most important. All other systems become less important. As individuals in a group, we express such a point of view.

Fig. 1-3. An individual may have one or more "blocks" that will keep him/her from participating in group activities.

But some people don't behave this way, you say. That's right, because some of us take each experience for whatever it is worth. In other words, personal values are unimportant. Also, entirely different values and goals, depending upon circumstances, seem to affect action more than specific group values and goals. People who operate this way are said to live by expediency.

So it isn't always simple to say that people solve their problems or accomplish their goals by participating in group activity through social systems.

And we are still individuals. Each one of us may have one or more "blocks" that keep us from participating in a group. At least eight blocks have been identified. These are:

1. *Fear*. The rest of you are smarter than I am . . . you will ridicule my ignorance . . . if I say something. I'll lose my status or position.

2. *Insecurity*. All of you do not like me. My accent is different. My hair is a disaster. My clothes are out of style. I don't have your education.

3. *Lack of knowledge about the group*. I didn't know about this group before this minute. I wonder what the group is up to. What are its goals? Are they the same as mine?

4. *Lack of time*. I shouldn't be here. I have work to do back in the office. I haven't time to attend this meeting.

5. *Lack of skill*. Joe has more training and experience in this than I, so why should I even try?

6. *Vested interests*. I don't have any children, so why should I vote for a new school building and increase my taxes?

7. *Group values*. If I go along with this program, I'll be thrown out of my church, and my family won't like it either.

8. *Group demands*. Seems like we're always calling a special meeting or drive.

At times we consciously are aware of these blocks, but often we aren't. If we're to be efficient group members, we need to analyze ourselves and try to overcome our blocks and frustrations. And we need to help others do the same.

ADJUSTING TO FRUSTRATIONS

Even when we're not aware of our blocks, we frequently adjust anyway. Here's how we do it.

- *Aggression*. When our ideas aren't accepted or are ignored, we may strike back. We refuse to go along with other ideas and may attack other motives. We may even turn against ourselves.

- *Compensation*. Since we can't talk well in a group, we offer to take the minutes, arrange for coffee breaks, or do other errands.

- *Rationalization*. We have a frustrating experience in leading a discussion. So we tell ourselves that the group members weren't trying and that discussions never get any place.

- *Identification*. Getting on Joe's side is best. Then we'll be right because he's usually right.

Fig. 1-4. At least nine ways in which an individual will adjust to frustration have been identified. Here are six of them.

- *Idealization.* We tell ourselves we really are good participants or leaders.
- *Projection.* We place the blame on a minor or nonexistent factor. Or, we blame someone else for our own admitted faults.
- *Displacement.* We transfer the feelings we have for one person or group to another. The boss was a tyrant today, so we take it out on our family tonight.
- *Regression.* We feel secure only in old and familiar situations. If we don't get our way, we sulk.

- *Conversion.* Inadequacy becomes a physical complaint. "I have a headache, so I can't accept the responsibility."

Now all this could appear to be a rather negative way of thinking of groups and the individuals in them. To the contrary, in our very lives there are many examples of how groups can be useful in communication activities. Our very structure of government depends on effective group activity. Groups give us all continuity, strength, resources, support, impact, decisions, fairness. These attributes help both the individual and all of us as a group.

◄ Groups Act Like People ►

All of us have wondered about the actions of migrating birds. In flocks, they swoop, change directions, gain altitude, and come to rest almost as if they were one.

In a way, a "group" of people has similar characteristics and habits. A group also has personality and methods of action much like those of an individual.

Since all of us deal with groups or participate in them, we need to be more familiar with their characteristics. Sociologists and psychologists call these characteristics "internal dynamics," or the forces or energies present within every group of people.

As you deal with groups, always remember that members are more productive if all of them feel they have access to important lines of communication, if they believe they have relevant information communicated to them, if they know and accept their role, and if they think they are communicated with on matters that affect them.

GROUP GOALS AND OBJECTIVES

Groups, like individuals, have goals and objectives that are often called the "aims," or "ends," of group action. These goals guide activities and serve as the basis for measuring success or failure.

Keep the following ideas in mind when establishing group goals or objectives.

1. Similarity of individual goals influences the degree to which group goals become established.

2. Accomplishments will be greater when goals are well defined.
3. Group goals can motivate and influence the behavior of group members. Members who most fully accept the group goals are most strongly motivated to help the group achieve its goals. Those who reject the group goals devote their attention to their own personal goals.
4. There is a close relationship between members' understanding of goals and their participation in the group.

TECHNIQUES FOR REACHING GOALS

It's one thing to agree on group goals and another to accomplish them. So there must be some means, or method, for getting the job done. Here are some other points to remember.

1. Groups that seriously consider means for accomplishing their goals are more productive than those that follow no rational process or those that always follow traditional patterns.
2. Means selected must consider the individual skills, interests, and motivations of the members. Additional factors would include forces from outside the group, such as prestige of the group or organization among others, the community value system, and any existing or likely competition.
3. Means should be selected with a broad understanding of possible alternate methods.

THE GROUP ATMOSPHERE

Some groups are a joy to work with; others are not. Much depends upon the "atmosphere" or mood of the group.

The atmosphere can be friendly and warm, with all members able to express their views without fear of ridicule or rejection. Or it can be cold, apathetic, and unfriendly, with members needing approval from an authority before expressing views.

Obviously, more work will be possible and more progress made when a group is friendly.

INTERNAL GROUP COMMUNICATION

Effective and efficient internal communication is essential if group members are to participate and help the group achieve its objectives. Through communication, members become better acquainted with each other, share ideas and experiences, express views and opinions, and approve or disapprove of group actions.

Communication breaks down (1) when it is "one-way" or dominated by a few; (2) when the majority do not have the same meanings for the word symbols being used; (3) when gestures, tones, and inflections of speech contradict the words; and (4) when members are unfamiliar with or misinterpret the interests, backgrounds, and personal ambitions of their colleagues.

Fig. 1-5. Communication breaks down when the majority don't have the same meanings for word symbols being used, and when members are unfamiliar with or misinterpret the interests, backgrounds, and personal ambitions of other group members.

PARTICIPATION BY MEMBERS

As you deal with groups, always remember that members are more productive if all of them feel they have access to important lines of communication; if they believe they have had relevant information communicated to them; if they know and accept their role; and if they think they are communicated with on matters that affect them.

You can observe these feelings by asking yourself the following questions:

● What was the *breadth* of participation? How many members took part?

● What was the *intensity*? How many times did various members enter into discussions?

● What was the *pattern*? Which people interacted with one another? When one person entered a discussion, was he/she usually followed by certain others? Which people monopolized the discussion? What role did the leader play?

● Were there any non-verbal expressions that showed more subtle meanings?

Generally, a group member is more satisfied with group action when he/she has a chance to participate, even though others may not completely agree with his/her viewpoint.

GROUPS HAVE STANDARDS

Every group, just like every individual, has definite standards that guide its operations and activities. These standards may be high or low. They cover items such as membership, participation, meeting procedures, decision making, and getting things done.

Once established, these standards determine each member's expectations of the group and each member's performance level. High standards encourage high expectations; low standards encourage low expectations. The circle is hard to break.

Standards must be realistic. They must not be higher than the group can attain, or there will be continual discouragement. And all members should understand them thoroughly once they are established.

Standards usually will be higher and more closely followed if members, rather than leaders or administrators, establish them.

GROUPS EXERCISE SOCIAL CONTROL

Through social control, a group obtains conformity from its members. This control may take the form of tangible rewards for members via group recognition, election to office, status assignment, or actual gifts. Some rewards are more intangible, such as acceptance by the group, a feeling of response, or a "pat on the back."

Social control can also take the form of punish-ment. Erring members may be ridiculed, ignored, re-jected, or even expelled. Methods of social control should be understood by all members, and they usu-ally are. Their influence depends upon the impor-tance each member attaches to belonging to the group.

◄ The Roles of Group Members ►

All members of a group play one or several roles as they participate in group discussions and activities. Usually a member plays a role that fits his/her per-sonality. But with guidance and practice, you can en-courage any group member to play a different role that will contribute to group productivity.

THE GROUP TASK ROLES

Roles of members are related to the task that the group wants to undertake. Such roles coordinate group effort and make it easier to find and solve problems. Here are some important roles.

- The *initiator-contributor* suggests new ideas or a different way of looking at the group problem or goal. His/her proposal may be in the form of a suggestion for a new group, a new goal, or a new definition of the problem. It may suggest a different way to handle or solve a problem.

- The *information seeker* asks for clarification of suggestions in terms of their factual adequacy, for authoritative information, and for back-ground about the problem.

- The *opinion seeker* wants clarification of values pertaining to the undertaking. This person may ask about values involved in a suggestion or in alternative suggestions.

- The *information giver* offers "authoritative" facts or generalizations or relates a personal experience to the problem.

- The *opinion giver* states a belief about a sugges-tion or an alternate. He/she emphasizes what the group's view of pertinent values should be and doesn't merely give relevant facts.

- The *elaborator* spells out suggestions in terms of examples or developed meanings and rationalizes prior to suggestions. He/she also tries to deduce how an idea would work if adopted.

- The *coordinator* shows or tries to clear up rela-tionships of ideas and suggestions. This person tries to pull ideas together or to coordinate ac-tivities of various members.

- The *orientor* defines positions with respect to group goals. He/she hears a summary of what has happened, points out departures from agreed-to directions, or questions directions.

- The *evaluator-critic* subjects accomplishments of the group to a standard or set of standards of group functioning. This person may evaluate or question practicality, logic, facts, or proce-dure.

- The *energizer* prods the group to action or de-cision and tries to stimulate or arouse greater or higher quality activity.

- The *procedural technician* speeds up group movement by handling routine tasks such as distributing materials, rearranging seating, running projectors.

- The *recorder* notes suggestions, group deci-sions, or results of discussion and serves as the group's "memory."

GROUP BUILDING AND MAINTENANCE ROLES

Members help the group work as a unit. They alter or maintain the group's way of thinking. They strengthen, regulate, and perpetuate the group.

- The *encourager* praises, agrees with, and ac-cepts contributions of others. He/she shows warmth and solidarity toward others. In various ways, the encourager expresses understanding and acceptance of suggestions, ideas, and points of view.

- The *harmonizer* mediates differences between other members. He/she attempts to reconcile disagreements and to relieve tension by jesting.

- The *compromiser* operates from within a conflict in which his/her own ideas or position is involved. He/she may yield status, admit errors, practice self-discipline, or "go halfway."

- The *gatekeeper-expediter* attempts to keep communication channels open by encouraging, regulating, or facilitating the participation of others. ("We don't have Joe's ideas yet.")

- The *standard setter* establishes standards for the group to achieve or to use in evaluating the quality of group processes.

- The *group observer-commentator* keeps records of various aspects of group dynamics. Based on such information, he/she evaluates the group's actions.

INDIVIDUAL ROLES THAT OFTEN HINDER THE GROUP

Group members play individual roles when they want to satisfy individual needs. Sometimes these needs are irrelevant to the group task. Sometimes they interfere with the work to be done.

- The *aggressor* may work in many ways: deflating others; expressing disapproval of values, acts, or feelings of others; attacking the group or the problem; joking aggressively; or showing envy of another's contribution by trying to take credit for it.

- The *blocker* maintains a negative attitude and is stubbornly resistant. He/she disagrees and opposes without or beyond reason any attempts to bring back an issue after the group has rejected or bypassed it.

- The *recognition seeker* creates self-attention. Boasts and reports of personal achievements

and unusual acts are part of a constant struggle to keep from being placed in an "inferior" position.

- The *playboy* displays a lack of involvement in the group's processes. He/she may be cynical or nonchalant or may engage in horseplay and other forms of "out-of-the-field" behavior.

- The *dominator* tries to assert authority or superiority by manipulating the group or certain members. He/she may try flattery, assert a superior status or right to attention, give directions authoritatively, or interrupt the contributions of others.

- The *special-interest pleader* speaks for the "small businessperson," the "grass roots," the "homemaker," or the "average wage earner." Usually, personal prejudices and biases are cloaked in the stereotype that best fits the need.

Remember, individuals can adopt several roles at one time. Some roles may seem to block group success, but all roles help to assure completion of group efforts in solving problems.

If individual roles seem to conflict with group needs, there may be several reasons: (1) a low level of skill among members, including the leader; (2) too much authoritarian leadership; (3) too much laissez faire leadership; (4) a low level of group maturity, discipline, and morale; and (5) a poorly chosen or defined group task.

Evaluation can also affect group productivity. Each member constantly evaluates his/her role, status, contribution, and feelings toward the group. Each evaluates other members in the same way to determine if those individuals' interests or needs are being met by the group. Each also compares the group with other groups. All these are factors that influence productivity.

◀ The Social Action Process ▶

Programs involving people don't just happen. And things don't get done just because someone or some group has an idea. The idea is reviewed and discussed with others; it receives status; it spreads and is evaluated. It is accepted by more and more people who offer to help; goals are established, plans are made, resources are mobilized, and finally an action program is launched and carried through—with possible revisions here and there.

Although the process isn't exactly the same for every situation, most social action follows a familiar pattern.

It's easier to understand and appreciate the social action process when it fits into a pattern with various phases identified and labeled. With better understanding of the steps involved in the process, an educational leader can make this process work in his/her favor.

ACTION IN A SOCIAL SYSTEM

Human behavior always takes place in an existing social system. That system can be a collection of people who have little contact with one another, such as a nation, a state, a county, or a community. Or it can be a group in which families or working staff members have considerable contact. Activities in a social action program may not involve the whole social system, but such activities are likely to be affected by the whole system.

A prior social situation always has a definite relationship to the program being planned or in progress. Previous experience can affect both the attitude toward and the success of the new venture.

Social action usually has its beginning when two or more people agree that a problem exists and that something should be done about it. Often this agreement comes from people who are "inside" the social system. (Parents with children in grade school agree that traffic signals should be erected at a street crossing.) Agreement of need, however, can come from people "outside" the system. (A businessperson and a minister agree that signals are needed at the school crossing.)

Fig. 1-6. Social action usually begins when two or more people agree that a problem exists and that something should be done about it.

THE "INITIATORS" ARE READY

Once an idea or a need is defined, a few individuals must express enough interest in the problem to do something about it. This group usually is small and can be regarded as *Initiating Set Number 1*. In the school crossing example, the set might be two or more parents, a homeowner near the crossing, and perhaps a teacher or two.

Quite likely, there may be three or four initiating sets. The parents would be one set, homeowners another, teachers a third, and so on.

LEGITIMIZING THE IDEA

In nearly every community or social system, there are certain individuals or groups that seem to have the right to pass judgment on new plans or ideas. They're called *legitimizers*.

It's not always easy to determine why or how a person becomes a legitimizer. It may be because of family status in the community, the official position held, reputation for past correct judgments, or any combination of reasons. Also, different people are legitimizers for different kinds of social programs.

Most of the time, legitimizers may be classed as "formal" or "informal." Formal ones usually hold some official administrative authority. Informal ones may hold no official rank at all, but the members of the community accept them as individuals with authority. Sometimes formal and informal groups may be legitimizers.

Ultimately, of course, the people themselves are the final legitimizers, but the formal and informal ones usually pass judgment before the idea or plan gets to the people.

GOING FROM LEGITIMATION TO DIFFUSION

Once an idea or a plan has been legitimized, it is ready to be spread. Here we find out if the people, or a relevant number of the people, will approve.

Just as there are initiating sets, there are different diffusion sets. In the school crossing example, the PTA might be one diffusion set. Another might be the community teachers organization. Still another might be the city council or the homeowners association.

People who initiate an idea often do not play a role in the diffusion process, for they may not have what the public demands of action leaders.

PLANNING FOR CONTINUAL EVALUATION

Throughout the process, each stage needs evaluation. This leads to decisions, planning, and action related to the next step.

Such evaluation might include these questions: Did we get the job done? How well did we do? How could we have done better? Why did we fail? What should we do next? How should we do it?

GETTING PEOPLE TO ACCEPT THE NEED

Once the diffusion set is established, people who will be affected by the action must define and accept the need as it applies to them. Getting this acceptance isn't easy, but there are a number of ways of accomplishing it. These include:

- *Through basic education.* Take a long-range approach; get people to appreciate needs they did not know existed; and present alternative methods of action.
- *Through committees.* Have people serve on program development committees to become familiar with the need for action.
- *Through surveys.* Lead people through a series of logical questions about a problem, and they are more likely to recognize the need for action.
- *Through comparison and competition.* Compare your situation with those of other towns, other schools, or other farms.

ESTABLISHING THE GOALS

Once there is a commitment to do something, positive and concrete action can be taken. Definite targets, or goals, can then be established with alternative methods for reaching them. We frequently argue more over the "how to do" than the "what to do." When the procedure has been outlined, the program can be launched and carried through the established plan of action. Every communication skill must be coordinated and related to the purpose—another step in continual evaluation—and each new move must be made in light of this continual evaluation.

◀ The Diffusion Process—How New Ideas ▶ Are Communicated

STEPS WHEN INDIVIDUALS ADOPT

When we adopt any new practice, we generally go through a mental process that may cover a considerable period of time. For easy identification, these steps, or stages, have been given functional labels. These are:

- *Awareness.* Before we can adopt a new idea or practice, we must know about it. We become aware that it exists even though we have few, if any, details.
- *Interest.* An idea may intrigue us, so we want to know more about it. What is it? How does it work? What will it accomplish?
- *Evaluation.* Here general interest is turned into personal interest. We mentally place the idea or practice in terms of our own situation. We may ask ourselves: How can I do it? Will it work in my case? What else might I do? Will it be easy? What will I get out of it?
- *Trial.* If the idea passes the mental evaluation, we're ready to try it. We may plant a few acres of a new variety or try a new sewing technique. We want to test whether the idea or practice will work for us.
- *Adoption.* If the trial is successful, we're ready to adopt the practice. This may mean large-scale use, continued use, or satisfaction with results. We may only accept the idea or be in and out of the practice in terms of costs, markets, climatic

conditions, etc., but we are basically satisfied with the idea.

WHEN GROUPS ADOPT

When local planners or other community leaders start with a problem and begin searching for alternatives, the pattern is different from the research center-to-farmer flow. The usual sequence is: problem, search for alternatives, trial, adoption.

The selected alternatives often are products of research when the problems need technical solutions. Frequently, the alternatives are ones suggested by the extension workers. Of course, many types of group problems can be solved merely by agreement and joint action. And some types of group decisions leave no room for non-adoption by individuals—the majority rules. But, where technology is involved, whether the process starts with awareness of a new practice or with a problem, the pathways to knowledge are similar.

WHO ADOPTS WHEN

Innovators. When farmers are evaluating a new idea, most of them want to talk it over with other farmers (see column 3 in Chart 1-1). A few, however, are innovators, who like to try new things first. Innovators make heavy use of research and expert sources of information and are frequently in direct contact with extension and experiment station personnel.

Early adopters. Close behind the innovators come the early adopters. They tend to have more education than later adopters and to be active in community affairs, read more newspapers and magazines than most people, and reach the trial stage quickly.

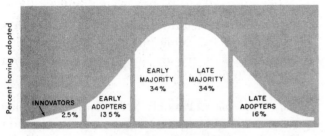

Fig. 1-7. One of the earlier adoption-diffusion studies showed the distribution of farmers among the five categories of adopters according to time of adoption.

Opinion leaders. Some of the innovators and early adopters are also opinion leaders. They are the ones other people seek most often for advice. Their opinions are valued highly. Different people are likely to be sought for opinions on different subjects.

The opinion leaders make ideal cooperators in trials and demonstrations of new technology in their communities. They are easily identified. Just ask a few individuals in a community whom people go to most often for advice on a particular subject.

Majority. Most people wait until the innovators and early adopters have tried a recommendation or practice, and proved it is practical. Other farmers have the most influence on these individuals' decisions to adopt.

Non-adopters. In most communities, a few individuals will be left who don't adopt a new practice. For some of these the decision may be a rational one, as in the case of an older couple who are cutting down on their operations and can't attempt anything new that involves more labor.

Recent studies allow for a discontinuance step. This may be because the idea or practice has been replaced with a better one or because the individuals became disenchanted. The practice may not have resulted in the expected advantage, or the advantage may have been small, and other people failed to adopt it or they made fun of it. Try to avoid having people discontinue use of new technology for the *wrong* reasons, such as the last one. Confirmation that other people are using the practice with success helps.

INFLUENCE OF COMPLEXITY

In general, the more complex and expensive the practice, the longer it will take for diffusion and adoption. For example, consider the following three types of changes with varying complexity:

1. *A simple change in materials.* It's easy to switch from one kind of fertilizer to another.
2. *An improved practice.* Switching from reliance on visual judgment to reliance on production records for culling cows requires additional knowledge, skill, and effort.
3. *An innovation.* Adoption of a newly designed solar heating unit for a farrowing house would require much study, investment, construction, and adaptation.

Fig. 1-8. Just as with new ideas, we go through the diffusion and adoption processes in accepting new commercial products or product innovations.

INFLUENCE OF FAMILIARITY

Familiarity with the concept involved can speed adoption. For example, it took about 14 years for wide adoption of hybrid corn because the concept was totally new. Radical adjustments in seed production and handling also were required.

But when hybrid grain sorghum was introduced, most farmers switched to it in three years or less. They were familiar with the hybrid vigor concept because of the corn hybrid results. Those in the corn production areas were quicker to change to hybrid grain sorghum than those who lived where grain sorghum was the predominant feed grain. The latter weren't as familiar with the hybrid revolution that had occurred in corn.

To extension workers this suggests that any time they can tie a new idea to a successful old one they can speed its adoption.

SUPPLYING RIGHT INFO AT RIGHT STAGE

At each stage in the adoption process, people have information needs that we can help them meet. We can influence quicker adoption by speeding the flow of information through the normal, preferred communication channels than we can by trying to bypass them. Note that mass media channels can be used to gain potential adopters' *awareness* and *interest*.

From there on, interpersonal channels become more important. In these later stages, the strategy in the use of mass media should be to help stimulate flow in the interpersonal channels. Announcement and follow-up news stories on tours, demonstrations, and educational meetings, for example, help put the topic on the neighborhood talk agenda by giving those who didn't attend enough knowledge to converse with and ask questions of those who did. The stories should identify participants so friends and neighbors can ask them about the event.

HOW THIS INFORMATION MIGHT BE USED

Let's trace the steps a new discovery might take in reaching a particular clientele group. Whether we are working with farmers, homemakers, youth, or any other group, the process is essentially the same. But in this illustration we are using a farm innovation for our example.

Assume the agricultural experiment station has developed a solar heating unit for confinement farrowing houses. Here are some typical steps you might take to hurry the process.

Assist innovators. Possibly a few innovators may have learned about solar hog houses before you begin a local information program. If not, perhaps you can contact a couple, inform them personally about the innovation, and offer some technical assistance—from yourself or another specialist.

Extra attention with these people is warranted because they serve an important function: They test new technology for others in the community, absorbing much of the risk. They will likely encounter problems in adapting the solar units to local conditions—supply problems, service problems, skill problems, etc. If you help them solve the problems

**Chart 1-1. Information Needs of Farmers
at Different Stages of Adoption**

Stage	(1) Functions	(2) Kind of Information Needed	(3) Preferred Sources
Awareness	Become informed	Notification	Mass media channels Other farmers Agencies
Interest	Become informed	More details	Mass media channels Other farmers Agencies

Evaluation	Self-persuasion
	(or legitimation)
		Will it work for me?
		Local trial consequences
		   • Economic
		   • Social
		Opinion of trusted others
		   • Farmers mostly
		Results elsewhere
			Trusted farmer friends
			Trusted others

Trial	Decision to use	Application
		   • How?
		   • How much?
		   • When?
			How-to publications
			Local dealers
			Self
			Neighbors

Adoption	Confirmation	Own results Experience of others	Own experience Other farmers

or learn how they solve them, you will be ready to help others convert to solar heat.

When one or more innovators have achieved success, enlist them in your educational effort, if possible, particularly the ones who are opinion leaders. Their involvement might include taking part in tours, demonstrations, meetings, and interviews with media reporters. The objective is to get other farmers talking with them and to get the innovation on the local farm talk agenda.

Fig. 1-9. At all stages of the adoption process, interaction with people we know and trust is an important part of the communication system.

Make others aware. While still working with innovators, start using radio, newspapers, television, and other means available to create awareness of the innovation among other farmers.

Supply details to expand discussion and interest. In the interest stage, farmers need to get more detailed information about what the innovation is, what its likely consequences will be, what it will do and will not do, how it will fit into their own plans, what it cost, how it compares to what they are already doing, what its economic and social consequences are—or, in general, those things that farmers want to know before accepting the innovation.

Radio broadcasts, personal conversations, and newspaper articles can be used to provide additional information. Potential adopters should be pointed in the direction of the local extension office and university publications.

Facilitate talk with successful users at evaluation stage. This is where the experience of innovators and others who have already tried the solar units will be most useful. Do whatever you can to get information about local trials and results spread around and talked about. Tours to farms where the solar units are in use and meetings and demonstrations where successful users are present to talk to other farmers will help. Report these events to the news media.

Supply publications at application stage. The information problem here is different. The farmer has already decided to use a solar hog house and needs simple, concrete information on how to build and operate one. Publications are an excellent way of providing this information. They can be read, filed, and read again.

The farmer may also want first-hand information about how to solve unresolved problems. You should be ready to refer farmers to whatever source could help—this could be another farmer, an extension engineer, or a supplier or builder.

Provide reinforcement information at adoption stage. Adoption is most dependent on results achieved up to now. The farmer's own experience is most important, but reinforcement from the success of others also helps. Thus, reporting the successes of others in the news media and other channels is helpful in sustaining adoption.

Listen to others. To be accurate in our own writing, thinking, and planning, it's essential that we pay attention to what people are saying and that we be perceptive to their thoughts.

Involve audiences. For true communication to take place, a two-way flow of questions and information must be established between the sources of information and those seeking the information. One of the best ways to provide this is to involve local people with a vested interest in a product or idea in planning an educational program to improve its production and promotion. This keeps the information flow related to local needs, sets up effective interpersonal communication channels in the community, and guarantees educators an audience of interested listeners and participants.

Most examples given have been in terms of agriculture. The principles, however, are the same for home economics, education, medicine, business, and other subjects.

◀ Planning Effective Communication ▶

Planning a communication program is something like taking a trip. We have to decide, first, whether the trip is necessary—what purpose it will serve. Then we have to decide *where* we're going before we consider routes or the kind of vehicle we might use to get there. In other words, we have to make a few decisions about goals.

Asking ourselves four questions will help.

1. What is the *need to be met* by the messages we are going to transmit?

2. What types of *communication goals* are we trying to reach? Teach skills? Inform people of a new method? Persuade people to accept a new belief or adopt a new technique?

3. What *obstacles* must we overcome? These can arise from our audience, our own purpose, and our limits in time, money, and facilities.

4. What *specific outcomes* do we have to obtain to get to the general goal we have in mind? This question requires that we be more specific about just what is to be communicated to whom, in what situation, and by what means.

Fig. 1-10. Several important decisions must be made early (and sometimes quickly) in the communication planning process.

The success of the program or campaign can depend largely on how well we answer these questions. Answers to one set of questions will depend partly on how we've answered the others.

Things don't always work that way, of course. Sometimes we can plan by checking where we are in a given situation, adjusting our approach as we move along. And sometimes we won't have time to ask ourselves the four questions, much less think out answers, before we start communicating. But still we should keep them in mind. Then if we're caught in an off-the-cuff situation, we can switch our approach, even as we start to communicate. We may decide, for instance, that the problem really isn't what it first seemed.

People may ask for information when what they really want is reassurance. Or, our purpose in giving them information may really be to persuade them to adopt some action. Therefore, we're not just answering a question.

Practice looking at situations and analyzing them in some detail, using these four questions. The experience will help you to react more quickly when you have an off-the-cuff experience. In other words, when time is short, you'll have a ready-made shortcut.

WHAT IS THE NEED TO BE MET?

Consider the origin of your problem. Did it come from "within"—from us? From the boss? From the organization? Or, did it come from "outside"—from the people you serve?

If origin is from within our organization, we may need to explain or report something or to persuade someone. In other words, we may have to arouse interest on someone's part.

If origin is from outside, it may be that interest already exists and the need is merely to get help or information.

Real need may not be apparent at first. For instance, people may not know exactly what information they want. We may have to help them.

Sometimes we may need to play down the main goal. Even when we originate the need ourselves, we may not want to make a frontal attack—especially when we're trying to persuade or motivate an audience.

DETERMINING GOALS

We must determine what to communicate in order to meet the needs we've identified.

Let's consider two kinds of goals: (1) teaching specific information or skills; (2) more general "educating" in terms of *influencing attitudes* that will lead to a given kind of action later on.

By grouping goals, we've already started to think of differences in kinds of messages needed to achieve them. Now we can really get specific.

If we're talking about Group 1 goals, we can attain teaching objectives by asking: What performance do. we want? What specific things need to be learned to get this performance?

Answers might go like this: We want to teach people to identify things.

Or, we want them to follow certain fixed procedures.

Or, we want them to know how to respond in different ways.

Then, again, maybe we want to teach some kind of final performance.

Or, maybe we're more concerned with a specific skill, such as using a typewriter, driving a car, or operating hand tools.

Each objective may require a different set of operating procedures.

But if we're talking about Group 2 goals—*communication for influencing*—our first problem is *getting attention*. That's not too difficult if an individual or a group has come to us for help. But if we originated the need, getting attention is always the first and sometimes the most difficult step.

The next step is likely to be the building of *favorable attitudes*. An important question arises: How specific is the proposed action and how soon is it to be taken? Our problem is close to building interest, but instead of overcoming apathy and getting attention, we may now have to try to eliminate some initial negative bias.

There's no sure rule to follow, but the way to overcome undesirable attitudes will become clearer the more we know our audience—know its biases.

Often it isn't the immediate effect that counts but rather the building of a foundation of attitudes or convictions on which later action can be based. Sometimes this will have to be done gradually.

With Group 2 goals, we'll also want to develop plans for an immediate commitment relating to specific behavior changes. If we build up a favorable feeling toward our message, we should provide a way for our audience to put it into action right away.

But chances are we'll want sustained behavior changes, attitudes, and skills that will last for a long time. Then we'll want to provide a means for a learning experience that will allow people to become more self-reliant. That's simply "helping people help themselves."

OBSTACLES TO REACHING GOALS

There probably are many ways to classify obstacles. Here are three:

1. *Personal obstacles*. These are found in both source and receiver. Included are items such as our knowledge, attitudes, communication skills, and socio-cultural background.

2. *Situation and resource obstacles*. These include limitations of budget, time, and place. Rarely do we have all the money, time, or facilities to do what we think would be best. So we must do the best we can with what we have, keeping in mind the size of our audience, the competition for its time and attention from others, and the adequacy of the materials we have available.

3. *Content obstacles*. These involve the nature of the material to be communicated, the skill to be learned, or the concept to be grasped. We need to ask ourselves: How new will this material be for the members of our audience? Have they ever heard of it before? What do they already know about it? How complex is the process? Is it like something the people are doing already? How much do they need to know to make the changes we want? How can we tell everything they need, briefly and concisely, without being boring? What will this mean to their own operations?

In other words, we must place ourselves in our receivers' shoes. That, of course, is more easily said than done.

DEFINING SPECIFIC OUTCOMES

For communicators, there's probably nothing better said than "start where people are." We can do this in terms of knowledge, attitudes, interests, abilities, and the like.

And we can ask ourselves again: What do we want people to do? How do we expect them to be different? What kinds of behavioral steps, in sequence, do we anticipate? What, in a sense, will be the first and last outcomes?

Now we can select the message content, communication code, treatment, and transmitting channels accordingly. We can stake out in advance certain subgoals or checkpoints to show that we are succeeding or failing and why.

All our actions must be taken in a framework of analysis and evaluation. The skills we apply to the situation are only a part of the task.

Now that you've planned, don't forget that you'll want to know how the plan worked—well, poorly, so-so. What was the effect on your audience?

Three possible effects are changes in knowledge, changes in attitudes, and changes in behavior. There can also be any combination of these items. If you've done a good job in planning, it should be clear which of the three effects you want. That also tells you which is important to evaluate.

Sometimes you can check or pre-test an idea. For instance, before you go to the expense of producing final materials, you can determine, by using other publications, whether people will pick up a publication with a colored cover more often than one with a black-and-white cover (behavior). If you're preparing a slide/tape presentation for 10-year-olds, you can determine if you've used the correct language by having four or five youngsters read the script (attitude).

Or, you may wait until after a magazine story is published and then go out to see who read it and got from the article what you intended (knowledge).

Actually, we evaluate all the time. We say things such as "This publication isn't very good," "That's a poor photograph," "The meeting was deadly." But these kinds of comments often only make the people involved feel bad. They don't explain how to overcome mistakes or what to do to make them right.

Systematic evaluation differs. We plan for it. We figure out what we would like to know about the effect of what we have produced. When we find that out, we can make changes in what we do the next time.

If we can't say what to do differently next time, evaluation isn't going to help us. So we need to think carefully about what it is we want to find out so that the answers we get will help us.

For example, suppose you find that readers don't understand a chart in a publication. That information alone doesn't tell you what to do to improve the chart or why it didn't do the job. But if you learn that the title didn't describe what was in the chart and there were too many trend lines for the reader to separate and understand easily, then you have knowledge that will help you make some concrete changes for improvement.

If you've followed all the steps to effective communication listed so far, you're well on your way to evaluating. You know the needs your communication effort is intended to meet. You've specified in planning whether the need comes from within or

outside. Evaluation can help you in communication planning by allowing you to check those needs.

Are they really needs for the intended audience? Just because someone comes to you and says that a certain group needs certain information . . . is that true? When it is important to check, you can use evaluation to go to that audience and find out.

Also, after you have produced the communication message or messages, you can check that audience again to see if the work you did actually met their needs.

If our communication goals have been Group 1 (teaching specific information or skills), we evaluate as we develop teaching materials to see if what we are developing will do the job. We also test at the end to see that our final product actually teaches people the information or skills we want them to have.

Group 2 goals (influencing attitudes) are tougher to get at. But the same ideas apply. Does what we put together communicate in a way that we can test or can see attitude change? We'll have a better chance of determining what happens and judging whether or not what we wanted to take place actually did if we can state those attitude changes in terms of behavior changes.

As far as overcoming the obstacles we envisioned when we first began planning our communication effort, we may now want to evaluate whether or not those obstacles actually appeared, and if they did,

Fig. 1-11. Specifics, such as communication channels used, as well as broad objectives, can be evaluated.

whether or not we overcame them adequately. We must have overcome the obstacles if our intended audience generally seems to have gotten our message and knows or is able to do whatever we wanted. Otherwise, we have to go back to some of the obstacles we anticipated—or discover new ones—if we're to explain why our efforts failed to whatever extent they did.

The more specific we were in naming outcomes for our communication effort, the easier it should be to evaluate and the surer we'll be in knowing success or failure. If our desired outcome was for people to understand better our industry, our system, or our organization, then we have a good chance of success, since almost any evidence can be accepted as success. But most of the time that will tell us nothing about what people really know or what they might do.

But suppose our outcome is for the audience to contact one of our field offices for a copy of a certain publication within two weeks of the diffusion of our message. No other action will do in our search for success. If enough people, in terms of how many or a given proportion of a known population, contact the office in the required time, we have succeeded. Otherwise, no. So we must go back to evaluate what

we did and when, or we must find out why people didn't do what we expected. Then, next time, we'll have a greater probability of success, for we learn from both success and failure.

You may want information about the audience, about the message, or about the channel—information that will help you before you prepare a message or afterward to find out what happens to that message. So evaluation is something you can do whenever you have questions, have choices to make, aren't certain what is best, or want to know the outcome of what you did—anywhere in the planning and production process.

Evaluate by whatever means you can. Find out which version of a publication people attending a field day tend to pick up the most. Ask a few members of a group what they learned from a certain speech. Conduct a full-fledged survey of a special-interest group—such as farmers—to find out what information and facts they need to be able to understand and use a new concept or program. The point is, gather information to help you.

And build evaluation into your job. Don't try to do it in your spare time. Show its value to your employer. Get him/her to expect it as part of your job.

◄ Let's Get Organized ►

Not long ago, a wise cartoonist drew a picture that received wide distribution. We'd be almost certain to find it in any office we visited.

The drawing shows two men, each relaxing in his office chair, each balancing a cup of coffee in his hand and stretching his legs out to some resting place. The caption says, "Tomorrow let's get organized."

We say that too when we're appointed to committees; when we start a new job and find that things aren't working so smoothly; when we're in a rush; or when we're too tired to take on another task.

For some reason, we think that if we're organized we can improve our efficiency. And well we might. The goal is to organize for "predictability," so that we can predict what we ourselves or what others will do under certain conditions—what their roles are likely to be.

So we make some rules. We have rules for operating our offices, our departments, our sections, or whatever we call the organizations for which we work.

Maybe a hierarchy develops. We make more rules. If we're not careful, we find that we, and others, react by doing just enough to get by. That's the result of a rigid organization.

FORMS OF GROUP STRUCTURE

All group structures assume certain shapes. There is *vertical structure*, in which the leader appears at the top and individuals of lesser rank are further down the list. There may be several levels at which some individuals will have equal rank but different roles. And roles may be quite well defined by placement on the chart. Such a structure is common to military organizations.

There is also *horizontal structure*. Everyone seems to have equal rank but different roles and tasks. Much of the work may be handled by committees. When subgroups meet for coordination, they will democratically select a leader. And leadership will usually change from time to time.

Some religious groups are organized horizontally, particularly among the leaders of the congregation. A combination horizontal-vertical structure also may exist where power is centralized in one office but work moves across the horizontal framework.

Structure also might be *chain-like*. It might be circular, so that suggestions, opinions, and decisions make a round of the various units within the overall structure. Often such an organization can't move with much speed.

Regardless of the formal organizational structure to which we belong, some members of the group will have certain shortcuts. If we're newcomers to a group, we'll overlook these at first, perhaps. Then again, if we're perceptive, perhaps we'll see these shortcuts rather quickly, but we may not find them so easy to use.

We try to match names to the organizational chart. We see who fits where in an informal structure. We might even see that the formal structure doesn't provide real control. In fact, the informal structure may really determine operations. For instance, say a head of a group has an in-law relationship with a legitimizer. Formally they have no connection, but they may pool ideas and values—and influence each other's decisions.

The better acquainted we become with the structure of our group or organization, the better we learn how to facilitate communication. If we're not aware of the structure, communication slackens.

STRUCTURE AND PURPOSE

Rules and regulations are the *technical* procedures for carrying out the objectives of an organization or a group. We might refer to those procedures as a manual of regulations or even as a *"Robert's Rules of Order."* The people who operate under certain procedures in an organization generally are referred to as the *bureaucracy*. But the point is, the bureaucracy is organized for a secondary purpose. The primary purpose is to achieve the goals of the organization. How goals are achieved is secondary.

Unfortunately, we too often forget that a public organization's purposes relate to people, not just to the organization or its legalities. Even if procedures are clear-cut, however, we are likely to create additional restrictions on communication. For instance, we may have wonderful communication skills and still fail to grasp the implications of scientific knowledge for a news report.

We may be skilled both in communication and in knowledge of all fields, yet be unaware of audience attitudes and perceptions of co-workers whom we represent. In other words, how we operate in any structure frequently depends mostly on us.

We—as bureaucrats—may be totally unaware of public attitudes and perceptions of our group. Because of the often changing structure of our group, we may perceive colleagues as having shifting values and perspectives when they may only be finding ways to communicate with policy makers. So we misperceive each other's actions.

Thus, important tasks to undertake within any public organization, group, or structure become the following:

1. *Define mission and role*. Discuss with others at all levels in our group our basic and ever-changing

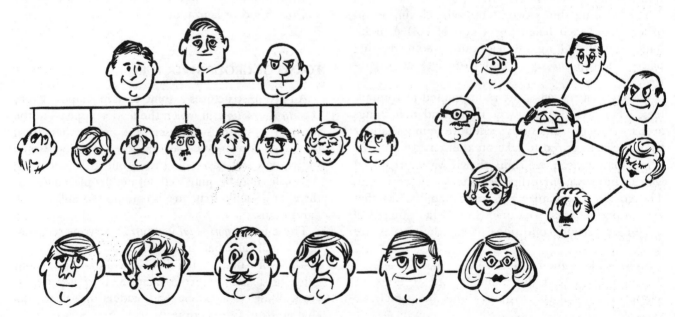

Fig. 1-12. Group structures can be vertical, horizontal, or chain-like, as illustrated here.

public commitments. These commitments are likely to be set by ourselves and for us by those whom we serve.

2. *Group purposes.* Assign specific roles, acknowledge internal interest groups, take advantage of status groups, and recognize beliefs and values within our organization. These aspects of structure affect the maintenance and change of policy decisions.

3. *Defend integrity.* Maintain our policy, our mission, our special capabilities, and our identity not for mere organizational survival but to continue our mission as long as it is needed.

4. *Limit internal conflict.* Win the consent of colleagues and others within the hierarchy of an organization, while continuing a balance of power that is appropriate to necessary production.

As communicators, we must be concerned with motivation. We want individuals to perform to their fullest potential within their assigned roles. We are building an atmosphere of learning, a process for change. By being organized and understanding our structure, we can better understand each other. Then we can all learn and, hopefully, reach the goals for which we aim.

◀ Leadership in Communication ▶

How often do we blame our leaders when something goes wrong? More than we'd like to admit, perhaps. But leadership does have much to do with effectiveness—especially in an organization, a committee, a meeting group, or any other work force.

Sometimes disagreements have a healthy result. More ideas can arise. But disagreements also often show that some members of the group do not have clear-cut notions of what is being attempted. Expectations are nil. In short, we do not have a clear direction. We haven't reconciled individual goals and group goals.

GOALS, GROUPS, AND LEADERS

We've already discussed some of the things on which group effectiveness depends. A lot also depends on how well the activities of different members are coordinated toward the common goals of the group—how skillfully the leadership manages the group and is accepted by it.

Among the many descriptions of leadership, two types are most commonly discussed—"democratic" and "authoritarian." Here though, we'll look at those two from a slightly different point of view.

One, the democratic approach, implies that *leadership is a property of a group.* The authoritarian approach shows leadership as a *characteristic of an individual.* The first can be called "participatory"; the second, "supervisory."

In organizations, most of the time we make leadership a group property. In fact, we're forced to do that because our organization—not an individual—

must show leadership in many, many activities. And all of us in the organization are leaders at one time or another.

Too often, however, we tend to make leadership synonymous only with individual prestige or with the holding of an office. If we do relate leadership to the performance of activities important to the group—our group—we sometimes tend to divorce ourselves from failures or mistakes. "They" did so-and-so, we say, instead of "we" did it.

We can think of our own leadership roles within our organization in an individual manner, however. We each possess certain personality characteristics—dominance, ego-control, aggressiveness, and ambition. We tend to pair off, to operate in subgroups, to form an informal structure aside from the formal group to which we belong.

As communicators, we can think of our leadership roles in terms of our functions: planning, decision-making, or coordinating. We can also add our many functions as individuals—executives, policy makers, experts, external group representatives, controllers of internal relationships, purveyors of rewards and punishments, and even ideologists, protectors, and scapegoats.

The questions then become: How well do others accept our single role of leader? In what terms do they think of us? How well do we understand other roles?

We need to realize that each of us is a leader of sorts. Others may think of us as "participatory" or as "supervisory" leaders contrary to our desires. But we can be either, if we're skillful. The major outcome is more effectiveness. As a skilled leader we can get more effectiveness mainly because we can make better use of minority opinions. A group with an unskilled leader never hears from the minority.

In an organization, whether we're participatory or supervisory leaders will depend upon our goals and our formal assignment. Even if we're elected or appointed to head a committee, there might be a slight favor toward being participatory leaders, because then it's generally easier to get expressions of opinions from all members of a group. Supervisory leadership sometimes squelches spontaneity and creativity, if the group thinks it's being manipulated by direction of authority.

Fig. 1-13. Leadership has much to do with group effectiveness, especially in an organization, a committee, a meeting group, or any other work force.

A leader may arise in different ways. Leaders may be appointed, be chosen by the group, or inherit the job. No matter; once a leader is acceptable to the group, he/she can give the group a feeling of responsibility by making each individual want to belong. Even an appointed leader can do this and establish acceptance by others who had nothing to do with the appointment. A leader can create a feeling of importance within each member by assigning roles important to the group. Then once the members of a group see that the group has a feeling of worth and acceptance toward each member, they usually will work eagerly and aggressively on the functions needed.

More Specific Leader Types

While prestige or status may be common to nearly all leadership, a leader really is merely one who performs any of a variety of jobs, which are popularly recognized as leadership jobs. The leader may provide the focus for the behavior of group members.

Or, the leader may merely be the most "popular" member. A good leader may do a job simply because it is expected, may perform better than he/she is able, or may do the job with such originality as to be recognized in the pages of history. Here are a few types, categorized by job:

- The *mobilizing or rallying leader* accelerates the work of the group. This type understands group members and has a way of looking at their ideas that makes sense and allows prediction of behavior.

- The *ingenuity leader* was chosen, probably because others anticipate a contribution to the group.

- The *defending leader* has a role primarily in relation to persons outside the group. A defending leader may have been chosen because he/she understands the policies of the group so well as to act in newly arisen situations just as the group would want.

- The *compromise leader* may be expected to play a role that may be of little value to the group other than to keep things rolling smoothly.

Who's the Best Leader?

The personal attributes of leaders often appear very simple and common. It's said that they seem to look at everything as if each moment were their last. They know how to listen and often take copious notes, or at least capture ideas by being alert and responsive. It seems that they try to understand first and judge later.

Generally, they welcome ideas and urge others to do their best thinking on a subject. They usually are open, sensitive, responsive, aware, and encouraging.

Few, if any, leaders waste time. They value it highly and use it skillfully. Perhaps without any certain listing, they tend to have definite goals, for leaders seem also to expend their energies toward solving problems that keep them from reaching goals. Obviously, they are highly self-confident, for they generally anticipate achievement and build on their strengths—asking clear and concise questions and organizing their approach to challenges so they can focus just on relevant tasks and problems.

No matter how we type leaders, the values of leaders are likely to vary according to whom we talk to. For instance, organization members in nonadministrative roles are likely to say leaders are best who (1) have more intimate social relationships with them, (2) are more sympathetic and indulgent in policies relating to supervision, and (3) pay little or no attention to formal status differences between themselves and others.

But members who hold supervisory or administrative positions in an organization seem to rank leaders highest who minimize interpersonal relations. They tend to associate "good" leadership with initiative, judgment, and formal organization of the group.

A major feature of a group, which affects its leadership and all other aspects of its functions, is the communication system or pattern available to it. It is the process by which one person influences another and is therefore basic to leadership. A person's position in a communication pattern determines largely both the assumption of group functions and the probablity of being perceived as a leader.

Sometimes these factors determine more than anything else who is or is not a good leader.

Photo/Art Credits

Bill Ballard, North Carolina State University: Figs. 1-5, 1-8, 1-9, 1-10, 1-11, 1-13.
North Central Regional Extension Publication No. 13: Fig. 1-7.
Third Edition: Figs. 1-1, 1-2, 1-3, 1-4, 1-6, 1-12.

2 THE ART OF GOOD WRITING

Writing is one of the oldest and one of the most important methods of communication. It is a means of presenting our thoughts, ideas, and information "of the moment," as well as our primary system of recording and preserving our thoughts for the future.

◄ Know Your Readers ►

Have a clear picture of the people you are writing for—age, sex, education, income, occupation, level of language spoken.

Consider the extent of the readers' knowledge—not only what they know but how well they know it, what they want to know, and what they need to know. Are they aware of the information you are offering, seeking more information, evaluating, trying the idea, or already using it?

Your writing success depends upon how well your readers accept what you tell them. To create reader acceptance, you must be aware of their interests, likes and dislikes, capacity for understanding, limitations, misapprehensions, and prejudices.

The reader fails to understand the writer when the writer fails to understand the reader. Remember that readers want to identify with everything they read. You can hold your readers' attention more easily if they feel that you are interested in them and their problems and that you convey this interest in your writing. You can't afford to be dull and boring because the competition for a reader's attention is too great.

Few people can resist good sales techniques. Good writers are good salespersons. They:

- Arouse the readers' interest.
- Stir the readers' imagination.
- Make readers feel the copy is aimed directly at them.
- Present believable claims.
- Give good reasons for action.

Fig. 2-1. Your success as a writer depends in part on how well you know the characteristics of your potential readers.

Prepared by **Geraldine Kessel,** press specialist, West Virginia University, with assistance from Gary Beall, communications specialist, University of California, and John W. Spaven, international communication consultant and former head, Office of Information, University of Vermont.

No substitute replaces planning, and a plan starts with a well-defined goal. You are not ready to write until you are able to state your purpose in one sentence. Then every word you write should bring you one step closer to this goal.

Good writing starts with an outline of the material you are going to present. Writing without an outline is like building without a plan. With an outline to guide you, you can avoid repetition and be certain you have included important points in spite of interruptions.

With your readers in mind, make a list of the main topics you will cover, and arrange them in the order in which you will write about them. Choose the order that seems most logical for your material. For example:

- *Chronological*—spring planting through harvest.

- *Step-by-step*—procedure for baking a cake.

- *Different categories*—chemical, mechanical, and biological weed control methods.

Under each main heading, list the related points.

◀ Write for Easy Reading ▶

Your readers are free to ignore your efforts any time they want to. And they will if what you write is not easy to read.

Here are some essentials for easy-to-read writing.

1. *Be conversational.* When you write as people talk, you have a style that readers are familiar with. (They don't mind if you end a sentence with a preposition.) Using common, ordinary words in a familiar way helps to convey the message. Don't become a slave to style. Be concerned first with getting your thoughts across to your readers. Use contractions when that's the natural way to say something but not when you want emphasis. Read your writing aloud. How does it sound? Smooth over the rough places and correct the grammar. Always remember you're writing to express your thoughts—not to impress your reader.

2. *Use short, easy-to-understand words.* When necessary you can use longer words if they carry your precise meaning. But don't forget that short, familiar words make the story easy to understand and hence easy to read. Rudolph Flesch has worked out a way to measure reading ease. It's based on the idea that the more syllables in a 100-word sample, the more difficult it is to understand. Here's a rough guide to measure your writing by: Easy reading—from 100 to 130 syllables per 100 words; standard reading—from 131 to 160 syllables; difficult reading—above 160 syllables per 100 words.

3. *Use personal words.* Words about people make writing more interesting. You'll find a gold mine in personal pronouns like *I, you, he, she, we,* and *they.* People feel you're really talking to them. Syllable counting may make your writing easy to read, but you defeat your purpose if it's also dull and impersonal. Use discretion, though, in putting personal pronouns in a straight news story.

4. *Use short, varied sentences.* In general, the shorter the sentence, the easier the reading. An average sentence length of 17 words is about right—but that's the *average.* Don't be afraid of stringing out some sentences if you can maintain a clear meaning throughout. And don't be afraid to bring the reader up short with a three- or four-worder. When sentence after sentence dutifully continues at 17 words, *rigor mortis* soon sets in.

5. *Use short paragraphs*. Block out one thought at a time so readers can catch their breath before going on to your next point. Good clear writing usually comes in neat little paragraphs of, say, three to five sentences each (with a one-sentence paragraph occasionally for emphasis or variety). Then the readers don't have to absorb a mishmash of thoughts.

Fig. 2-2. People prefer to read material that is less complex than they are capable of understanding. Short words, sentences, and paragraphs help.

6. *Put down your thoughts in logical order*. Good writing comes from clear thinking. You're asking too much of your readers if they have to unscramble your thoughts. Here's a system that will fit many of your stories: (a) State the main point in one sentence. (b) Tell why that point is important. (c) List all other facts and figures the readers should have on this particular subject.

7. *Check these points*.

Mixed tenses: "The farmer seeded oats in 3 fields and has plowed the back 40." (Delete "has.")

"He is writing for farmers and slanted his information to meet their needs." (*Rewrite*: "He writes for farmers and slants his information to meet their needs.")

Dangling modifiers: "Born in Alaska, her writing experiences began on the local newspaper." (*Rewrite*: "Born in Alaska, she began her writing experiences on. . . .")

"To get the most out of a long film, the seats must be comfortable." (*Rewrite*: ". . . film, you must have a comfortable seat.")

Nonagreement: "The use of radio, television, and news releases insure the success of the program." (*Rewrite*: "The use of . . . insures . . .")

"These kind of activities train young people for citizenship tomorrow." (*Rewrite*: "This kind . . . trains . . ." or "These kinds . . . train . . .")

Careless repetition (repeat words for emphasis): "Market gardeners annually produce tons of fresh garden produce." (*Rewrite*: "Market gardeners produce tons of fresh fruits and vegetables each year.")

Redundancy: "The consumer evaluated the value of the purchase." (Delete "the value of.")

"The beautiful lakes and streams contributed to the beauty of the state." (Delete "beautiful.")

Abstract nouns: "The school personnel failed to agree on the disciplinary action needed." (*Rewrite*: "The teachers disagreed on the discipline needed.")

Empty nouns: "The poor quality of the soil was the main cause of repeated crop failures." (*Rewrite*: "The poor soil caused repeated crop failures.")

"Attacks by the insect may be cyclical in nature." (*Rewrite*: "Attacks by the insect may be cyclical," or "The insect may attack in cycles.")

Hedging: "Apparently the tests seemed to indicate that the cattle disease evidently had been spread from farm to farm on the farmers' shoes." (*Rewrite*: "The tests showed the farmers had spread the cattle disease from farm to farm on their shoes.")

Circling (useless "thing is thing"): "The project all of the members selected was the home beautification project." (*Rewrite*: "All the members selected the home beautification project.")

Mixed construction or faulty parallelism: "The couple walked slowly, caterpillars falling on their heads and clung to their shoulders." (*Rewrite*: ". . . and clinging to their shoulders.")

Words are your tools. Toss out those that are shopworn, but avoid the unfamiliar and complicated.

Words will not think for you; they will only express what is on your mind. If an idea is hazy in your mind, words can't clear it up.

When you find yourself having trouble putting something into words, stop trying to write and start trying to think. Once you really know what you want to say, you won't have any trouble finding the right words.

People prefer to read material that is less complex than they are capable of understanding. Your readers are more likely to quit reading than to reach for a dictionary.

Though most words are workers, some just take up space. The many forms of "there is" often are among the useless words. For example:

"There has been an increase in the amount of milk consumed by teenagers."

Let's eliminate the non-working words:

"More milk is being consumed by teenagers."

This is an improvement, but let's get some action into it:

"Teenagers are drinking more milk."

This sentence is short, simple, easy to read, and every word in it is a worker.

- **Use the short word:** "is"—*not* "exists."

- **Use the simple word:** "ate"—*not* "consumed."

- **Use the personal word:** "you"—*not* "one."

- **Use the active word:** "It bit me"—*not* "I was bitten by it."

- **Use the colorful word:** "big as a basketball"—*not* "very large."

- **Use the specific word:** "110°F."—*not* "quite hot."

- **Use the familiar word:** "beekeeping"—*not* "apiculture."

- **Write simply, clearly, briefly.**

- **Know your readers.** Keep your readers' image before you as you write.

- **Have a plan.** Make an outline and follow it so your readers can follow you.

- **Make a rough draft.** Get your ideas down; leave the polishing for later.

- **Choose words carefully.** Use simple, familiar, personal words.

- **Say it simply.** It's not enough to write so that you may be understood; you must use language that can't be misunderstood. Don't allow anything irrelevant to intrude.

- **Don't present unfamiliar ideas without adequate explanation.**

- **Don't waste your readers' time;** they are busy too. Don't elaborate the obvious.

- **Punctuate for clarity.** Make your point with periods.

- **Use capitals sparingly.** Save them for special occasions.

- **Be your own editor.** Be ruthless; other people will be.

- **Send up a trial balloon.** Ask a peer to review your article.

Wordy sentences, too many prepositional phrases, and passive verbs produce hard-to-read, ineffective writing.

Look at this 20-word sentence: "The discussions become quite heated at times, which is an indication that the members themselves direct the policies of their organization."

You can cut the length in half by saying: "Occasional heated discussions show that the members direct their organization's policies."

Often the word "there" is not needed. For example: "There are more than 20 cultural projects from which one can choose to satisfy his/her interests, needs, and home conditions." (22 words)

Why not say: "A member can choose from more than 20 cultural projects to satisfy his/her interests, needs, and home conditions." (19 words)

Wordy: "There are many weaknesses in writing which can be detected if it is carefully reread."

Rewrite: "You can detect many writing weaknesses if you carefully reread your copy."

ELIMINATE PREPOSITIONAL PHRASES

Prepositions are useful, but your writing will be more effective if you use them sparingly.

Excessive prepositional phrases: "This contest is sponsored locally *by* the Central County Service Company *in* cooperation *with* the County Extension Office." (18 words, 3 prepositional phrases)

No prepositional phrases: "The Central County Service Company and the County Extension Office sponsor the contest." (13 words, no prepositions)

USE ACTIVE VERBS

Passive verbs have a place in our language, but most writers use them more than is necessary.

A verb is *active* when it shows that the subject acts. A verb is *passive* when the subject of the verb is acted upon.

Active verbs put action into your writing. For example, "The chairperson called for a motion to adjourn" is stronger than "A motion to adjourn was called for by the chairperson."

In the following examples, active verbs strengthen the sentences:

Weak: "A wide range of learning experiences is provided under the guidance of a volunteer 4-H leader."

Stronger: "Volunteer 4-H leaders guide members through a wide range of learning experiences."

Weak: "The farm of William Portage was also approved as a farm plan."

Stronger: "The board also approved William Portage's farm plan."

You can spot passive voice simply by watching for *"am," "is," "are," "was," "were,"* and *"been."* If they occupy space as auxiliary verbs ("are featured," "was liked") in a sentence, you are looking at passive voice. (Sometimes you must use forms of *"be"* as principal or linking verbs. Example: "Live verbs *are* in the active voice.")

WEED OUT REDUNDANT WORDS

Instead of . . .	Say . . .
very latest	latest
absolutely complete	complete
necessary requirements	requirements
basic fundamentals	fundamentals
cooperate together	cooperate
the consensus of opinion is	the consensus is
ask the question	ask
for a period of two weeks	for two weeks
factual information	facts
refer back	refer
check up on	check
end up	end
inside of	inside
all of	all
fold up	fold
paid out	paid
later on	later
over with	over
the necessary funds	the money
destroyed by fire	burned
during the time	while
completely destroyed	destroyed
as a general rule	as a rule

Cut out compound prepositions and conjunctions such as "insomuch as," "due to the fact that," "to such a degree as," and "in view of the fact that." "Because" or "since" says the same thing.

USE SHORT, SIMPLE VERBS

Instead of . . .	Say . . .
excavate	dig
contribute	give
interrogate	question
purchase	buy
was the recipient of	received
assist	help

Instead of . . .	Say . . .
obtain	get
hold a conference	meet
take action	act
terminate	end
inquire	ask
conceal	hide
attempt	try
possess	own

SAY IT SIMPLY

If you want to get your message off the printed page and into your readers' minds, say it simply so that they will remember it easily. Do you recognize these famous words?

"In this case I have undertaken the journey here for the purpose of interring of the deceased. From this point of view I do not, however, propose putting anything on record in so far as praise is concerned."

No? Then, surely you remember the edited version:

"I come to bury Caesar, not to praise him."

Photo/Art Credit

Bill Ballard, North Carolina State University: Figs. 2-1, 2-2.

3 WRITING FOR NEWSPAPERS AND MAGAZINES

Newspapers and magazines are businesses first and public services second. Publishers must meet their payrolls, pay for supplies, and earn a return on investments. After that, they can indulge their desires for public service.

Income comes from the paper's circulation—the number of copies sold each day. Roughly speaking, this determines advertising and circulation revenues. Despite the appeal of comics, crossword puzzles, and columnists, circulation depends on news.

◀ What Makes News? ▶

News is a precise, accurate, and current report of the facts surrounding an event. Editors weigh one news story against another when deciding which to publish. They generally base their decisions on a set of standards called "news value." The following items are often called the elements, or components, of news.

1. *Timeliness.* While weekly papers will use last week's news, daily papers generally don't use yesterday's news. On the other hand, education stories need not be an account of something that has just happened. Stories might be seasonal, for instance. A guide is to time a story to appear in the paper when the majority of the readers are in need of the information or at a time when most readers will be either doing a job or planning for it.

2. *Localness, or proximity.* Some small-paper editors prefer to use only stories that originated within their circulation areas, while others seek out news from outside the local areas if it is relevant or important to their readers. Larger papers attempt to cover "all the news," but still give the local stories a big advantage in getting into their columns. Also consider the "psychological" closeness of an Ann Landers–type column.

3. *Prominence, or importance.* The crucial word is "big"—what the event means to the audience, how big the event is in the community, or the effects of the project. Also, "prominence" refers to the status, or standing, of a particular individual. The definition of what is "big" depends on the circulation area also. For a small weekly, a big name in the weekly's own town usually is "big." For a regional daily, a big name would have to be known by many readers throughout the region.

4. *Conflict and consequence.* Struggles of people against their environment, or against one another, individually or in groups, make for interesting reading. What will happen as a result of the event?

5. *Progress, or change.* There is much interest in changes brought about by humankind. Many of

Prepared by **Geraldine Kessel,** press specialist, West Virginia University, with assistance from James Shaner, information specialist, University of Missouri, and Ralph Mills, photographer, North Carolina State University.

these changes have been beneficial, resulting in new technology, sophisticated production methods, modern equipment, better living conditions, and improved human relations. However, many have also been detrimental, resulting in soil erosion, energy wastefulness, and water and air pollution.

6. *Unusualness.* Anything out of the ordinary, often of little importance or consequence, makes the news. Rare, odd, or sometimes unforseen ideas, events, or situations qualify.

7. *Human interest.* Ideas, events, or situations which touch human emotions classify as "human interest." Human interest events may only arouse curiosity, but they may also incite anger, or elicit fear, joy, sadness, compassion, or other feelings. People are interested in other people—especially children and elderly persons—and in animals.

If we used the preceding list as a scorecard for judging a news article, we would be concerned with the number of news components found in the article as well as the amount of each component in one story.

On the other hand, editors know that readers' tastes differ widely, so they seek variety. They know that only a few readers cover the entire paper; most read only what interests them. Readers select their articles by scanning the headlines and possibly the first paragraph of each story.

Editors, therefore, cram as many articles into the papers as possible. Since space is limited, this means short articles with the hot air removed. The most important facts must come first, where they can be seen by the scanning readers. Otherwise, they go right by.

Fig. 3-1. The newspaper business is changing. Today editors and reporters on the larger newspapers use computers and cathode ray tube (CRT) screens instead of the traditional typewriter. You should be aware of the physical facilities in your local newspaper offices.

There is great variety among newspapers. In circulation they range from metropolitian dailies with several hundred thousand circulation to the local weekly printing less than a thousand copies.

The chances of hitting the big ones are slim. On the other hand, the small papers (printed one, two, or three times per week) are usually looking for stories, especially local ones. If you maintain good relations with the editors, they will likely print almost anything you submit to them. It is not unusual for the representatives of government agencies, local clubs, and other groups to "make the front page" on these small local newspapers.

Fig. 3-2. Study the newspapers you will be writing for. Know the kinds of stories they want and how they handle the stories they print.

Visit editors at work. But, because they are very busy at certain hours, it's both courteous and wise to phone for an appointment.

These visits will do several things. They will make you a living person instead of a faceless name to the editors or one of their staff. It will stamp you as a person who wants to learn. It will give your material an edge over material of equal merit from an unknown.

Beyond that, it will give you a chance to exchange views and sharpen your thinking. You may be able to tell the editors about story ideas and possibilities they had not thought about. Editors can give you ideas on how to use the newspaper to better reach your clientele.

Editors can tell you what they prefer and are most likely to use, when they will need it, how long it should be, when they want it written out (almost always), and when they would prefer a phone call (important news, close to the deadline).

Visit your editors often. And get them involved in your programs, wherever and whenever possible.

The mechanics of printing also shape newspaper style. Without plenty of extra work, no one can predict exactly how much typewritten copy will fit in the assigned space when reproduced. Editors make informed guesses. Then they tell the shop people to cut the story from the bottom, if necessary, to make it fit. The shop does this routinely.

That's why every regular newspaper article, as distinguished from a special feature, has its main facts at the beginning of the article. And each article is written so that it can be chopped from the bottom, paragraph by paragraph, and still make sense. This affects your writing. You can't make a vague statement at the beginning and explain it at the end. You must be solid from the start.

WRITING THE LEAD

A popular formula sums up the first paragraph, called the lead. It's the five W's of *who, what, where, when,* and *why.* When vital to understanding, add the *how* to the five W's.

Tell your story in terms of a local person or persons, the *who* in your article. The *what* is the most significant thing they did or said. Tell *why* they did it and *where* and *when.* These facts go in the first one or two sentences.

TESTING THE ARTICLE

Give your lead two tests.

1. Is every word vital? Will the readers be confused if any part of the lead is taken out? (If the answer is yes, the lead meets the test of tight writing.)

2. If the readers scan just the lead, will they learn the main facts? Will they have a capsule summary of what you want to tell them? (If the answer is yes, you've written a newspaper lead.)

After the lead, expand your article, adding facts in order of importance. Repeat the names of the persons you're talking about frequently to keep the readers oriented.

Before mailing the article, give it the imaginary scissors test: Will it make sense if it is cut from the bottom, paragraph by paragraph? Your answer should be yes in every case, because you have no guarantee that the entire article will be printed.

THAT LOCAL ANGLE

To make your material attractive, stress its timeliness. Try to tell the story in terms of local people and local scenes.

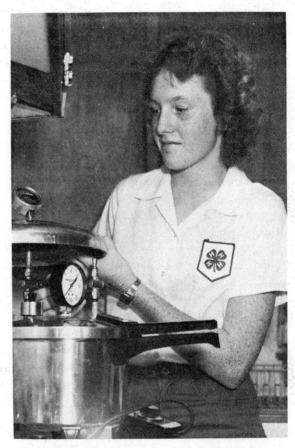

Fig. 3-3. Often a general story, or one received from a state headquarters or outside source, can be localized by using a photograph of a local resident. Here the story was about food conservation. This photograph, with the person identified, localized the story and helped to get it used in the paper.

The best method is the success story. You might, for example, recommend a farming practice by quoting a local farmer who used the method last year. Describe the method used, outline the results, and give the plans for the new season.

In the course of your story you have (1) attracted reader attention by writing about a neighbor, (2) listed recommendations, and (3) proved that the recommendation got results. The same technique can be used for any regular recommendations.

TIME AND PEOPLE

Each article, whether it's about a cake, a cow, or a camp, should be timely and have local interest. If it has several local names and comments, it's on target. Nothing is ever certain in the newspaper business, but such an article would be as close to a sure thing as technique can make it.

Making sure of timeliness and local interest takes a little extra effort, particularly for beginners. The story that pleases the editor will make the greatest impact on the readers. And that's what educational writing is all about.

NEWS AND EDUCATION

Your material falls into two broad classifications: news and education (sometimes called subject matter articles). It's helpful to know the difference.

News generally deals with a particular event at a particular time. The election of officers, the discovery of a new insect in the county, and the happenings of a meeting are examples of news. The common thread is the importance of the time element. Something occurs at a particular time.

Editors stress the newness of news. It must be handled quickly, as soon after the event as possible. Daily papers prefer to publish news no later than the day after it happened. Publication the same day is the ideal. Similarly, weekly editors frown on news that is more than one week old.

If the news is important enough, editors will accept telephoned reports. Most educational news is not in this category. That means a typewritten article should be on the editor's desk well before the deadline of the next edition.

But in most educational articles, the time is seasonal rather than critical. For example, an insecticide

should be used in the next two weeks. Lawns must be started by the end of the month. Milking machines need regular check-ups.

As with all newspaper material, educational articles should be timely. However, educational material can be held briefly. Editors welcome it when it's good, but it takes second place to solid news. It's generally good to indicate a release date—or the latest date the story will be timely.

ACCURACY HEADS THE LIST

The safest way to get a name straight is to ask the person to spell it. The next best authority is the phone book. If in doubt about a fact, ask your source. Nothing seems quite as permanent as an error made in print.

Guesswork is out for spelling names and addresses and for giving facts and quotations.

TYPING THE ARTICLE

Your article should have an identifying label, called a "slug," in the upper left-hand corner. If the article runs more than a page, write "more" at the bottom of the first page; then repeat the slug at the top of each succeeding page. At the end of the article, write "-30-," for example. The "more" tells the editor to look for another page. The slug identifies the story. The "-30-" tells the editor it has ended.

Put your name, address, home and office phone numbers on the article. This permits the editor to reach you for more information, if desired.

Material should be typewritten, double-spaced, on one side of the paper only. End all paragraphs on the sheet on which they start.

◄ News Stories, Features, and Columns ►

Of the several types of newspaper stories, columns are always written by the correspondent or educational agency representative.

Advance and follow-up news stories are usually submitted to the newspaper office in written form, although at times they may be phoned in and the newspaper may send a reporter and/or a photographer out to cover them.

Feature stories may be written by the educational agency representative. Or the representative may act as a liaison on a story, arranging for the interview between source and newspaper reporter. The representative is then present at the interview, as a specialist in the subject matter being written about, to insure that the reporter asks all the pertinent questions.

THE ADVANCE STORY

A good advance story takes work and imagination. Too often, the writer simply gives time, place, speaker, and topic and relies on the readers to supply the interest. On the other hand, the effective advance story has a built-in interest factor that attracts the readers. These two leads demonstrate the difference.

Lead 1. A meeting on livestock outlook will be held Wednesday at 8:00 p.m. at Gordon High School. All farmers are invited to attend.

Lead 2. Skidding hog prices and uncertain cattle markets may bring a record turnout for the Extension Service meeting at 8:00 p.m. Wednesday (July 23) at Gordon High School.

Both leads give the bare facts. But only the second one gives the why behind the meeting. And if the meeting doesn't have a solid why angle, there shouldn't be a meeting.

The time, place, and other facts help the readers reach the meeting place. But unless you convince them that they should go, they won't leave home. Tie your story to the readers' interests. To do this, be specific and let the readers know why the meeting is important to them.

A word of caution: We can get overenthusiastic when describing an event we have planned. But, regardless of the phrases we use, there should always be a "source" for the story. In other words, quote whoever has made the statements or judgments used in the story.

THE FOLLOW-UP STORY

The educational payoff is in the follow-up story. You can remind those who attended of the highlights of the meeting, reinforcing the learning process. (Strangely, we all enjoy reading about something we've seen ourselves, even though we presumably know about it already.)

Summarize and report the important points coming out of the meeting, because you have a simple obligation to the editor to follow through on coverage. By printing your advance story, the editor has accepted your statement that the event has importance. If this is true, then it deserves coverage.

This article also reaches the people who didn't attend. Almost invariably, they form the majority of the community. So, sift out what is important (in terms of reader interest) and make that the lead. Take good notes and verify, with the speaker, anything that may not be clear. Don't start with how many people attended the meeting, follow that by saying "an interesting time was had by all," and close by listing in a very general way the names of the speakers and what they covered.

In writing the follow-up, think again about your readers. What's their situation? What do they want to know? What should they learn that they haven't even thought about? How would the meeting have helped them?

The answers help you select the major items from the evening's events. Then ask yourself which one was the most important. That's your lead. Consider keying on one or two points in a self-interest lead. For example, "Boone County farmers can learn . . ."

But remember that stale news is worse than no news. After all, who wants a Christmas present in April? Get the article to the editor's desk before the next edition of the paper. This takes planning, but it's worthwhile.

FEATURE STORIES

Special articles, called feature stories, almost defy description. They frequently have a strong time element, but not always. They usually deal with people, but not always. Mainly, they are interesting because they have a touch of the unusual, the unknown, or the unrecognized.

Fig. 3-4. Always be on the lookout for human interest angles. Unusual things people do, children, oldsters, pretty girls, and animals have a high degree of built-in human interest.

Most people have a gold mine of features. They can tell of the achievements of local people in easy language. Readers are encouraged by the successes of their neighbors and are likely to imitate their methods. Features are excellent teaching tools.

Guiding the Interview

Professional feature writers look for people who are really interested in their work or hobbies. Such subjects are willing, even eager, to talk. Their enthusiasm shines through their words. The feature comes alive.

The interview makes or breaks a feature. The writer must get the subject talking—draw the person out. Questions built around "how" or "why" do this far better than queries that can be answered by yes or no. Discuss items of maximum interest to the person being interviewed.

After this area has been exhausted, turn to more sensitive areas, such as prices or losses. Wind up with biographical questions to give a relaxed finish to the interview.

Collecting Direct Quotes

Take notes or use a tape recorder during the interview. This is the only way to be accurate. It may make the person being interviewed a bit more nervous, but the results are worth it. After all, the subject is aware he/she is being interviewed; notes only confirm the fact. But, as a practical matter, it's probably best to get the interview rolling before the pencil comes out.

If occasional verbatim quotations are written down, use them within quotation marks in the feature. Good verbatim quotes dress up even the most modest feature, although no one really knows why this is so. Quotes are powerful.

Loose Style

In writing the feature, use modified newspaper style. Put the most interesting item first. But unlike the article lead, the feature opening is not necessarily a summary. Place the best items near the top, to hold the reader, but at a more leisurely pace. You're telling an interesting story to your neighbors, not giving a frantic news bulletin telegraph style.

Although features are written to be published without cutting, surprise endings are dangerous. Newspaper emergencies do happen, and features occasionally are shortened. So it's well to avoid that final switch in meaning. Good feature writers look

for a final tag line with a wallop. But they avoid final paragraphs that are essential to the sense of the story.

It's hard to explain a cockeyed story to angry people if the "April Fool" at the end was never printed.

PERSONAL COLUMNS

A personal column is something of a journalistic hybrid—part article and part feature.

A column is an island of opinion in a sea of factual material. The author can give personal views with no restraints—except the laws of libel, the dictates of good taste, and the willingness of the editor to print them.

Fig. 3-5. A standing head calls attention to a regular column. Such a head usually includes the title of the column and the name and title of the writer. In addition, a photograph of the writer, a drawing, or the symbol or emblem of the writer's agency makes an even better heading.

Although editors have the final say on everything printed, their main requirement of a column is that it attract readers. A column needs a following to hold its place in the paper.

Variety of Items

A writer can build readership by sprinkling the column with local names. Nothing attracts attention faster. Use names naturally, in connection with timely advice and reports.

Most successful columnists use several short items in each column rather than one long topic. Lengthy treatment may appeal to one group, but others will find nothing of interest. True, their turn may come several weeks later. But by then they may be ex-readers of the column. Only by including many different topics can the author appeal to the widest possible group.

A column also gives the writer a chance to promote meetings, point with pride, thank cooperators, and comment on events.

Gathering Material

A weekly column can become a chore. The well runs dry. Yet editors feel, and rightly so, that a weekly column should run every week, not at the writer's convenience.

Old hands have found that a folder for ideas helps. A reference can be made to a magazine article and dropped in the folder. Questions from the public always suggest items, if they're remembered. Again, the folder. Observations from the field visits or meetings beef up the reserves.

Form the habit of jotting down impressions at the first opportunity. Too much happens to rely on memory at the end of the week.

Some people who are able public speakers bog down before a typewriter. Such people may find it easier to dictate from notes. Since most columns have a conversational, informal tone, dictation can do wonders. Of course, edit the transcript before final typing and mailing.

Fact and Opinion

A final word about the difference between a fact and an opinion: In newspaper terms, a fact is primarily sense information—something we saw, heard, touched. Opinion involves a judgment, regardless of whether it's right or wrong.

For example, "It was a good day" is opinion. "The day was a cloudless 78°F." is factual. On the other hand, "John Smith said, 'It was a good day'" is a fact if we heard him say it. We're vouching for the quotation, not the accuracy of his opinion.

The source of all opinion must be identified in newspapers.

NEWS RELEASES FOR WEEKLIES SCORECARD

Possible Points

1. CONTENT 50

   News Value: Is material timely, newsworthy, and educational? Does it cover the subject? Are there evidences of in-depth reporting?

2. STYLE 45

   Readable: Sentences of varied length? Direct verbs? Easy to understand? Good grammar?

   Well-Organized: Is material expressed in a clear, logical order? Does copy flow easily?

   Treatment: Fresh, interpretive approach? Facts vivid and meaningful? Any human interest? Technical phrases explained?

3. APPEARANCE 5

   Is copy neat, clearly reproduced, and well-spaced on the page? Is format functional in getting stories printed? Is source of copy easily identified by the newspaper editor?

   TOTAL SCORE 100

Fig. 3-6. This news story scorecard from the annual contest conducted by Agricultural Communicators in Education highlights the relevant points of a good news story for weekly or non-daily papers.

◀ Fillers and Photographs ▶

Leaf through any newspaper, and you are likely to see small items at the bottom of news columns, usually separated from the stories about them by a single line. This item is called a "filler." It will usually have no relationship at all to the story in the column above it, or on either side.

Newspaper editors don't like to leave unfilled space in their newspapers. But they can't stretch the type. We said if a story runs too long to fit the available space, the editor clips lines off the bottom of the story. If the story fails to fill up the available space, he/she reaches up on the shelf for a filler.

> In 1979, American farmers harvested more acres than any time in the past 30 years and used a record amount of fertilizer to boost farm production to an all-time high. In the process, they used record-high acreage for exports and supplied more persons per farm workers than ever before. Moreover, they did all this with fewer tractors, corn pickers and shellers and pickup balers, even though total hours used for farm work hit an all-time low.

> The average Japanese worker spends 20 percent of his income for food. It takes the typical worker 32 minutes to earn a pound of butter.

> Consumption of chicken in 1979, at 51.5 pounds (ready-to-cook weight) was up from 47.5 pounds in 1978 and record high for the fourth straight year and consumption of turkey, at 10.1 pounds, was also up from the record highs of 9.3 pounds for 1977 and 1978.

> Beef consumption in 1979, at 79.6 pounds (retail weight), was off from 88.8 pounds in 1978 and the lowest since 1967. Consumption of veal, at 1.6 pounds, was off from 2.5 pounds in 1978 and the lowest since 1973, while lamb and mutton consumption, at 1.3 pounds, was down from 1.5 pounds in 1978 and the lowest on record.

> Boneless sirloin steak on January 6, at $2.90 a pound in Washington, was going for $4.48 a pound in The Hague, $5.50 in Brussels, $5.60 in London, $6.20 in Copenhagen, $7.70 in Stockholm. And, are you ready for this? $19.03 per pound in Tokyo.

> Fifty percent of a worker's income in India goes to buy food. The typical worker works 46 minutes to earn a pound of bread, compared to just four minutes for an American.

Fig. 3-7. Almost every edition of every newspaper needs a filler or two to fill up the columns. Some fillers from you can get your messages in this space. Above are some agricultural samples from several newspapers.

Where do fillers come from? From newspaper syndication services, from commercial concerns, from information offices, and from you.

Every now and then, send your editor a page of 6 to 10 fillers—facts about your area of work. He/she will keep them on hand, and when a filler is needed, he/she may use one of yours, instead of some fact about the Grand Canyon, a long ago sporting event, or the world's largest person.

PHOTOGRAPHS NEEDED

Probably every survey of newspaper editors to determine their needs indicates that they would like to have more photographs. Specifically, they will ask for photographs to accompany the articles you prepare for them.

Also, photo stories can be created from one or more photographs. Photos with a strong human interest angle are especially well received. Human interest is hard to define, but activities featuring animals, children, pretty girls, and elderly persons are generally given a high human interest value.

Photographs can make some of the best follow-up coverage. "Candids" that catch people involved in what they're doing are the workhorse of photojournalism.

WHAT THE EDITOR LIKES

"Will this interest my readers?" Each time your

Fig. 3-8. The winter wood supply piles up as Kevin Dailey throws a chunk away from the buzz saw. Elijah Dailey and Jim Clark (right), all of Route 1, Mercer, Missouri, prepare to saw another piece from the white oak pole they are holding. This photo is a good example of one that can stand alone. It tells a complete story, with local people, a timely subject, and a lot of human interest.

editor looks at a picture, whether it be a news shot of storm damage or a feature photo of a 4-H member with his/her sewing project, it must pass this simple test. If his/her editorial judgment votes yes, he/she will take a second look.

"Does this picture tell the story?" The editor must now decide whether the picture is clear and to the point. Is the presentation—through choice of viewpoint, selection of elements, and compositional emphasis—such that the average reader will readily grasp the meaning.

"Will this picture reproduce?" It must satisfy the editor technically. The print must be clear and sharply focused, with good contrast. The picture will be screened in engraving, which means it will be made up of tiny dots. These dots are transposed in printing to rough newsprint paper. Both steps tend to lose quality, so you can readily understand why the editor needs the best possible quality in the beginning.

"Is the information adequate?" No editor will accept a picture without all information necessary to write the caption. The newsperson's stock in trade is information. He/she must know who's in the picture, what it's all about, where it happened, and why it's important.

NEWS AND FEATURE PICTURES

The news picture is an effervescent thing. Today, it is a hot potato; tomorrow it will be a cold cucumber. Your job is to get it to the paper while it's still hot.

The feature picture, on the other hand, is a slow burner. It may never be as hot as the news shot, but it may be more leisurely taken and published with less dispatch.

The news picture is the sudden catastrophe, the newly elected officers, the first of a kind, the current event.

The feature picture is a personality, a job well

Fig. 3-9. Get as much action into each news photograph as possible, as long as the action is consistent with the subject or activity being pictured.

done, a pretty scene, a how-to-do-it task, a project report.

Both types of pictures may be used with written stories.

Presentation of News Pictures

The 8 × 10 or 5 × 7 glossy prints are accepted as standard in most newspaper offices.

Captions should be typed double-spaced on a separate sheet. The caption sheet may be pasted or taped on the back of a picture.

Pictures should be delivered in a manila envelope. If they are sent through the mail, they should be placed between two corrugated board stiffeners.

And in picture taking, as in writing stories, you should talk to the editor about his/her needs, special instructions, and other information you might get that will enhance your success in getting stories and pictures in the paper.

◄ Writing for Magazines ►

Pick your magazine; learn what its editor is trying to accomplish. No matter how well you write, you will not sell (or place) articles unless you write with a specific publication in mind. As in selling any product, if your first prospect doesn't buy, change your approach with the next.

Knowing your subject is not enough; you must also know your audience. This includes the readers of the target publication and its editor. So become familiar with articles in that publication. Obviously this does not mean that the editor would not consider a certain departure, but start with knowledge of the magazine's typical articles: length, type of information, apparent interest and education of the readers. Note extent to which articles are supported by pictures and writing style.

QUERYING THE EDITOR

Should you query the editor? Generally, yes. Query by letter, normally, but don't hesitate to turn to the telephone or to a personal visit.

Although the query is principally to locate an editor who shows interest in your topic, or approach, the querying can save time in writing. For example: You have the idea, the facts, and the hope of a 2,000-word article. To your query the editor replies, "We probably can use it if within 600 words with such and such a slant." You are thus saved the struggle of producing a 2,000-word article which will not be accepted.

You can query a second editor, but *only* if you clearly have no commitment with the first editor.

There are times, however, when you can market the same material after it has been published. But, you'll need a fresh approach, probably some additional or updated facts, and perhaps different pictures.

And regardless of the rule, hard experience has proved that sometimes an editor will accept an article in hand when he/she has or would have rejected it in the idea stage, especially on a query from an unknown writer. So break the query rule if you feel the urge to write first, and don't worry about being unknown the first try. Isn't everybody?

EDITOR'S HELPER

Perhaps you spot a situation that you think would be right for a certain magazine, but you have little confidence in your writing ability, or you are just too busy at the moment. Don't hesitate; editors can rewrite, you know. So select a publication and explain your idea to the editor. Tell why you think it would be well received by readers. Tell the editor what you could do to help. A magazine staff writer might do most of the work.

Editors like these special tips. Sometimes they will pay for such tips. And, as a bonus, this practice sharpens your ability to see a story situation, and it improves your contacts with editors.

SUBJECT MATTER

You don't need to be a subject matter expert to produce an informative, authoritative article. When you are aware of trends, you can spot developments that emphasize or clarify the trends. For example, the introduction of "challenge" or "lead" feeding to dairy cows a few years ago opened pages in a

---

MAGAZINE STORY SCORECARD

	Possible Points
1. CONTENT	40

Informative Value: Is material newsworthy and educational? Does it cover the subject? Are there evidences of in-depth reporting? Was copy provided with sufficient lead time?

2. STYLE	55

Readable: Sentences of varied length? Direct verbs? Easy to understand? Good grammar?

Well-Organized: Is material expressed in a clear, logical order? Does copy flow easily?

Treatment: Fresh, interpretive approach? Facts vivid and meaningful? Any human interest? Technical phrases explained?

Illustrations: Are photographs, graphs, charts, and other aids clear and concise? Do they strengthen the presentation?

3. APPEARANCE	5

Is copy neat, clearly reproduced, and well-spaced on the page? Is source of copy easily identified by the editor?

TOTAL SCORE	100

Fig. 3-10. This Agricultural Communicators in Education scorecard highlights the relevant points of a good magazine story.

wide range of farm papers for several articles per magazine.

PREPARATION AND ORGANIZATION

The feature article requires thought, investigation, and verification. Compared to writing a news feature, you will gather more facts, opinions, and quotations than you can use. Then you will select or combine them for a more meaningful article.

You will organize your material more thoughtfully. (This is not to suggest that a news story is not carefully organized, but that in a feature, a wider range of treatment and interpretation is possible; and your choice within this range could make or break the article.)

You will develop (or think) a story outline. Opportunities for dramatic leads are abundant, and you will demonstrate your artistic polish by revising and rewriting your article several times. In the process,

you'll eliminate clichés, overlong sentences, unsupported claims, extraneous facts or opinions, excess enthusiasm or emotion.

SOME DON'TS

Don't avoid figures and statistics. Their omission may leave an article with no real meaning. But do get them correct. And state them simply. "Nearly three out of four farmers" reads better than "72.3 percent of the farmers."

And don't get carried away with the thoughts that articles in mass media must be general. The readers have a specific interest in the subject presented, or they wouldn't be reading it anyway.

Your feature article must convey a message that is clear to the readers. Whether or not they agree is of no concern at this point. If they know what you wrote and what it means, you have produced a good feature article.

◀ Newspaper and Magazine Style Rules ▶

How many times have you sat down to write a news story or magazine article only to be faced with decisions about proper style usage?

The following rules, adapted from recent Associated Press and United Press International stylebooks, may help you. They were selected to answer some of the most common questions.

Academic degree and courtesy titles. If it's necessary to establish credentials, avoid an abbreviation. Make it "a bachelor's degree," "a master's," "a doctorate in psychology." Use "Ph.D.," "LL.D.," etc., when many listings would make the preferred form cumbersome. Set abbreviations off by commas.

Capitalize and spell out formal titles like "professor," "chairperson," "dean" *before* a name; lowercase elsewhere. However, it's "history Professor Joe Doe" or "department Chairperson Richard Roe."

Lowercase academic departments, except for proper nouns or adjectives: "English department," "Romance languages department." But capitalize names of schools and colleges and formal or official names like "Coal Research Bureau."

Follow this style for faculty, staff, and administrators:

First reference: full name and position or title (no academic or professional degrees noted).

Second reference to those with earned doctorates: "Dr. Jones."

Third and later reference to those with doctorates: (men) last name only or pronoun; (women) "Dr. Jones" or pronoun.

Second and later references to those without earned doctorates: last name only for men; last name plus "Mrs.," "Miss," or "Ms.," for women.

Use "Mr." only in "Mr. and Mrs." Use "Miss," "Mrs.," or "Ms." in first reference, except in cases of married women who request that they be referred to by their husbands' names; otherwise "Emily Jones" first, then "Mrs. Jones." Use "Ms." in second reference to women who prefer it or whose marital status you don't know.

Caps and lowercase. Capitalize (1) formal titles *only* when used before names (including "president"); (2) brand names: "Frisbee," "Vaseline," "Thermos," "Band-Aid"; (3) "EST" and "EDT" (time); (4) "pill" when referring to the (birth control) Pill; (5) "legislature" when referring to a specific one; (6) "papers" if given in quotes; (7) "Army," "Navy," etc.; (8) "X-ray." In the names of diseases, capitalize the name of the person and lowercase the disease: "Hodgkin's disease."

Lowercase acting as in "acting President John

Doe"; lowercase "a.m." and "p.m." It is *Time* magazine, but *Harper's Magazine*, since "Magazine" is part of the formal title.

Capitalize proper names of livestock, animals, fowl, etc.: "Percheron," "Hereford," "Yorkshire." Lowercase the kinds: "whiteface," "bantam," "terrier," "horse."

Spellings, plurals, parentheses. It is "adviser" ("advisory"), "aesthetic," "kidnapped," "transferred" (despite rule not to double final consonant in two-syllable words accented on first syllable), "lifestyle," "monthlong," "pocket-size" (not "sized"), "OK" (not "okay"). "Stadiums," not "stadia," "memorandums," not "memoranda," are plurals, but "symposia" is plural of "symposium" and "agenda" is singular, while "agendas" is plural.

Number can be plural in cases like "A number of plans have been considered." Add "s" to pluralize numbers, without apostrophe: "5s and 10s," "1980s."

Use parentheses sparingly; they jar the reader. Commas or dashes are preferred. But use parentheses for location inserted in a proper name: "Miami (Ohio) University."

Numbers. General rule is to spell out "one" through "nine," to use numerals for "10" and above. Use numerals for ages of persons and for percentages (1 percent). Note: AP-UPI stylebooks call for "percent"

as one word: "10 percent to 12 percent," and for the use of the "$": "$12 million to $14 million."

Use numerals for proportions: "2 parts powder, 5 parts water." For ratios, use "2-to-1, 6-to-5."

Abbreviate the month when the day of the month is used; otherwise spell it out: "Jan. 24, 1982," but "January 1982."

Miscellaneous. Use traditional abbreviations for states, not the two-letter ones from the postal list.

A person may graduate from a school; it's not necessary to say "he/she *was* graduated."

Put in quotes most names of books (newspapers generally don't set italics), but not scriptures (Bible, Koran, etc.) or reference works: dictionaries, encyclopedias, etc.

Use "vs." rather than "versus"; "All-America" (player), not "All-American."

If distinction is needed when referring to the dateline community of a story, repeat the name.

Close up space between initials: "R.L. Stevenson." Make it "5- or 6-year-old."

Semicolons go outside unless they are part of the quoted matter (unlike commas and periods).

"Then Mayor Earl McCartney" and "wind chill factor" are not hyphenated. But, use "T-shirt," not "tee shirt."

"Youth" refers to young people, age 13 through 17.

Photo/Art Credits

ACE Critique and Awards Program: Figs. 3-6, 3-10.
Bill Ballard, North Carolina State University: Figs. 3-5, 3-7.
Duane Dailey, University of Missouri: Fig. 3-8.
Vellie Matthews, North Carolina State University: Fig. 3-1.
Jerry Rodgers, North Carolina State University: Figs. 3-2, 3-3, 3-4, 3-9.

4 HOMEMADE PUBLICATIONS

Why write a publication?

Maybe you need to give specific instructions for a job. Or perhaps you need more details than you can give through the news media.

The answer is simple. Write a publication. You can get out a homemade publication quickly and cheaply. Studies have shown that readership and use are high when a publication is sent in response to a request.

If you've planned ahead and your publication is already available, then you can mention it in your newspaper story or on your radio or TV program. Your audience can read the publication at its convenience.

◀ Getting Started ▶

What do you want to say? To whom? Get a mental picture of your audience and the information needed before starting to write.

Is your purpose to inform, report, or persuade? State your objective. Then make an outline of items you want to cover. Organize the topics; then write. What's the first thing the audience should know or the first step the audience should take?

Remember, short publications cost less and have higher readership. Keep your sentences and paragraphs short. Use short, everyday words instead of those that have three or more syllables.

Tailor your publication to a specific audience—gardeners, beef cattle producers, turf grass managers, newlyweds, parents of young children.

You might consider printing pesticide information separately since recommendations change often. You wouldn't want to discard your entire supply of publications because the information was outdated.

◀ Visuals ▶

You may need some illustrations—drawings, charts, graphs, or pictures—to "dress up" your publication and help get your message across.

Get your photographs, especially seasonal ones, ahead of time. Black-and-white film gives the best prints for publication. Although black-and-white

Prepared by *Jimmy Tart,* senior publications editor, North Carolina State University.

prints can be made from color slides, they often appear dull or out of focus.

Line drawings can be reproduced faster with less expense than photographs (halftones). Look through other publications and collect sample drawings that you can use or adapt. Be sure to observe all copyright restrictions. If you can't draw, trace photographs or make "stick" figures by using lines and circles.

The Agricultural Communicators in Education (ACE) organization has compiled an 8½- × 11-inch book, with hundreds of ready-to-use drawings. The ACE *Clip Art Book Number 5* is published and sold by The Interstate Printers & Publishers, Inc., Jackson at Van Buren, Danville, Illinois 61832-0594.

Use charts or graphs only if you can make them simple. You'll still need to summarize the information in narrative form. A ruler or a straightedge and a lettering guide or "instant letters" will make your work look professional. Instant lettering comes in different sizes on transparent gummed sheets. Line up each letter and rub. It's transferred instantly. If you make a mistake, scrape it off.

◀ Production ▶

Your printing method will be determined by time, money, number of copies needed, and equipment available. Usually the least expensive is the mimeograph.

The Multilith, a small offset press, provides better quality printing than the mimeograph machine. You can draw and type on offset masters and go directly to press without the cost for extra negatives. Offset masters are designed to print a certain number of copies. Be sure to use those that will not "break down" before all the copies needed have been printed.

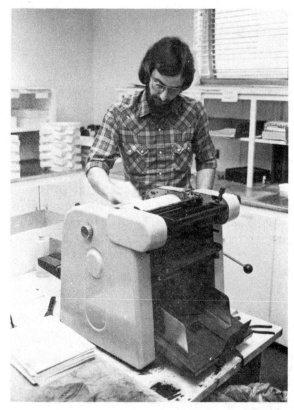

Fig. 4-1. The "duplicating center" has replaced the "mimeograph room" at many places. Small, tabletop offset duplicators will use paper masters directly from the typewriter or plates made on a platemaker.

Fig. 4-2. The small offset press will provide sophisticated printing, of one or more colors, and is efficient for printing single sheets and small publications to be folded.

If you have access to an offset mastermaker (platemaker), you can type, draw, and paste up on plain white paper. Text and illustrations can be glued or waxed down, and your offset master can be made from the paste-up.

Instant, or while-you-wait, printers can provide excellent results at low cost. There are generally two master-making processes—photo-direct and electrostatic masters. The photo-direct process will usually give better quality at a slightly higher cost, but both

Fig. 4-4. Publications come in a variety of finished forms. Here a small, automated collator assembles the printed single sheets and staples them together—the simplest bindery work. Discuss the different finished forms with your printer.

Fig. 4-3. A medium-sized offset press, found in many small commercial print shops, will print as many as eight pages (6- × 9-inch page size) at one pass through the press.

processes give better quality than copy machines or duplicators.

An office copy machine might be used to print your publication if it gives good quality copies. If you have to buy special copier paper, costs will probably be higher than those for mimeograph or offset printing.

Photographs will slow your printing process and add to the cost. If you're using photos, use a nonreproducing pencil to mark locations and sizes.

◄ Spacing and Margins ►

The 1250 Model Multilith will take paper 10 × 14 inches, but copy, centered on the page, must not exceed 9½ × 13 inches. The larger press will take 11- × 17-inch paper, but the copy area is limited to 10½ × 16½ inches, centered.

The margin is the space between your typed area and the edge of the page. Always leave the widest margin at the bottom. Your side and top margins may be alike, or the top margin may be wider. Single pages that will not have holes punched should have equal side margins. Facing pages usually have narrower inside (gutter) margins.

For a four-page leaflet with a finished size of 5½ × 8½ inches, use 8½- × 11-inch paper, folded. Treat the front and back covers (pages 1 and 4) as single pages and the inside pages (pages 2 and 3) as facing pages. Margins for pages 2 and 3 might be: ½ inch inside (gutter), ¾ inch top, ½ inch outside, and 1 inch bottom. The result will be a typed page 4½ × 6¾ inches—the same shape as the whole page.

◄ Reduction of Copy ►

To produce a page of copy that is 4½ × 6¾ inches, type (as close as possible) 5 inches across the page and 7½ inches down the page. Then reduce 10 percent, and the result will be readable type, but more copy per page.

In laying out facing pages for the 10 percent re-

```
9 POINT
ABCDEFGHIJKLMNOPQRSTUVWXYZ
abcdefghijklmnopqrstuvwxyz
1234567890

10 POINT
ABCDEFGHIJKLMNOPQRSTUVWXYZ
abcdefghijklmnopqrstuvwxyz
1234567890

11 POINT
ABCDEFGHIJKLMNOPQRSTUVWXYZ
abcdefghijklmnopqrstuvwxyz
1234567890

12 POINT
ABCDEFGHIJKLMNOPQRSTUVWXYZ
abcdefghijklmnopqrstuvwxyz
1234567890

14 POINT
ABCDEFGHIJKLMNOPQRSTUV
abcdefghijklmnopqrstuvwxyz
1234567890
```

Fig. 4-5. Type is selected on the basis of size. Here is a popular typeface showing comparative sizes, with the height of the letters specified in "points" (1 inch = 72 points).

duction, leave about 1⅛ inches between them (for the gutter margins). This space will reduce to 1 inch, giving outside margins (left and right) of ½ inch.

A line drawing will add interest to a page of typing. But white space can also be used to produce an eye-catching layout. Use three-character indentions for paragraphs, rather than the usual five-character indentions.

If your finished size will be 8½ × 11 inches, run two columns of type as shown on this page.

Use ALL CAPITALS and *underlining* sparingly.

Box (draw lines around) an important point to get attention.

Large dots (bullets) can be made by filling in the lowercase "o." Periods and hyphens can also be used to draw attention.

Various screens (lines, dots, patterns) are available in sheets to help you achieve halftone appearance and texture. Lay a sheet over a line drawing, press lightly, and use a razor or graphic arts (X-acto) knife to cut around the area to be screened. Lift off the rest of the sheet and burnish (lightly rub) the drawing with something strong and smooth, such as a putty knife.

If copy is to be reduced, do not use a very fine screen.

Communications
Methods and
Techniques

**Communications
Methods and
Techniques**

Fig. 4-6. Type is selected on density or line thickness, and is often designated as "light," "medium," or "bold." Here light and bold types are compared. There are other special types, such as italic.

◄ Layout ►

Keep the cover or the first page simple and uncluttered. Put the author's name on the back page or the inside cover (page 2).

Use a consistent scheme for heads throughout your publication. A major head that falls at the bottom of a page should have at least four lines of typing under it. Put at least two lines of type under a minor head.

Do not put a widow (short line at the end of a paragraph) by itself at the top of a new column or page. Carry at least one full line with it.

Unless your equipment will justify type, don't attempt to make lines come out evenly. Inserting additional spaces between words requires more work and makes your finished product more difficult to read.

Don't divide space equally on the page. Use variety to get informal balance. Make a dummy and use different shapes and sizes of copy, drawings, and pictures to find a pleasing layout.

Other layout principles are contrast and repetition. Obtain contrast by changing size, tone, texture, and direction. Repeating a line, curve, shape, or texture also produces interesting designs. Repeat a cover picture, drawing, or theme on the inside pages for effectiveness.

◄ Paper and Ink ►

Two decisions that must be made early in the process of getting something printed are the selection of paper and the selection of ink.

Black ink on white paper is probably the standard of the printing trade. However, you can usually choose other paper and ink colors with some but not a great deal of additional expense.

Papers are sold and labeled by weight. In mimeograph papers, onion skin, 13-, 16-, 20-, and 24-pound weights are usually readily available. For printing on both sides of the sheet, 20-pound weight is recommended. If the paper is to be printed on only one side, 16-pound is quite satisfactory. The 13-pound weight is usually used when it is necessary to hold down mailing cost or to reduce the bulk of a particular job.

Offset papers are also labeled by weight. Fifty-, 60-, and 70-pound weights are the most commonly used.

At the economical end of the line is newsprint; at the expensive end is a wide variety of fancy paper stocks and finishes. One way to dress up a mimeograph or offset duplicated publication is with a heavier cover stock, in color, with the title in large display type and with perhaps a photograph or drawing.

In printing terminology, a single color means one color of ink is used (black or some other color). Process color is used to get full color into a publication. This process is very expensive.

Present printing technology permits the use of several colors or tones from a single ink—through screens and similar devices—which gives the effect of more than one color. Also, colored paper is an economical way to dress up a publication.

Talk papers and inks with your printer. Buying inks and papers is similar to buying paint in a paint store. From the swatch books a printer has on hand, you can select—from dozens of choices—the paper stock and ink color(s) that you desire.

Photo/Art Credits

William L. Carpenter, North Carolina State University: Figs. 4-5, 4-6.
Vellie Matthews, North Carolina State University: Figs. 4-1, 4-2, 4-3, 4-4.

5 DIRECT MAIL

◄ Why Use Direct Mail? ►

Industry spends more than $4 billion a year (one out of every six advertising dollars) on direct mail advertising. The nation's fastest growing advertising medium, direct mail ranks second in volume only to newspapers among major media. Industry invests more money each year in direct mail than in radio, television, or magazine advertising.

● Two out of every three magazine subscriptions in the United States are sold by mail.

● An auto manufacturer invited millions—by mail—to win a prize by visiting dealers. And 38 percent of the recipients showed up.

● Six letters were sent to 550 firms to promote a resort hotel as a sales meeting site. They produced 80 bookings and 12,000 room reservations.

● A razor blade firm complemented mass media advertising with mail samples of a new stainless steel blade. This helped put the firm's blade in the sales lead.

● A political party sent 10 fund-raising letters, each focused on a different issue. Returns indicated one issue standing out as most urgent in voters' minds, and it became a key campaign topic.

● In a primary campaign, a 94,000 mailing brought in 11,000 pledges to write in a candidate's name. A follow-up asked the 11,000 to sign up other voters. And the final result was a 12,000-vote victory margin.

WHAT IS DIRECT MAIL?

The term "direct mail" means many things to many people. When we ask individuals if they use direct mail, they may reply that they send out a monthly newsletter. To question them further, we might ask if they ever send post cards to one or more people to tell them when they are coming to their area, to remind them of a meeting, or for any other reason. Their response might be "Sure, but that isn't direct mail." They think of direct mail only in terms of a newsletter to a special audience on a fairly regular basis.

Actually a post card, even to one person, is a form of direct mail. Why? Because it is a message directed to a specific audience to accomplish a specific purpose. This is the one determining factor. The size of the audience doesn't affect whether or not the message is direct mail.

"Direct mail" may also be defined as the "paratrooper" of communication. It can be directed to a specific audience to accomplish a specific job.

It takes teamwork to win battles. This is also true of communications. Radio, TV, newspapers, magazines, direct mail, and other communication methods are all fitted to do certain jobs best. And, just as paratroopers can't win a battle alone, direct mail can't do the communications job alone. When all methods are fitted together—when each is used for its particular job—together they can help to accomplish an effective job of communications.

You have to select the communications tool or

Prepared by **John W. Spaven,** international communication consultant and former head, Office of Information, University of Vermont.

method which is best suited for the job. If you have a message for a large audience, perhaps one of the mass media is your best channel. But if you want to pinpoint an audience, direct mail may be the logical tool to use.

WHAT MAKES GOOD DIRECT MAIL?

Three things make good direct mail: a good mailing list, a good idea, and a good approach in presenting the idea.

A good mailing list should be based on audience interest and should be accurate, complete, and up-to-date.

A good idea is one the audience wants, needs, or can use when the message is sent.

With a good approach, copy and format relate to the audience members' interests, needs, or wants. Copy is in the readers' language—in terms they understand. And the mailing has an "easy-on-the-eye" layout.

A PERSONAL CONTACT

Now let's look at some of the jobs for which direct mail is best fitted. Research has shown that mass media are most effective in creating awareness of, and interest in, an idea. The same studies have shown that personal contact is one of the most effective ways of getting people to try out or to adopt an idea.

That's where direct mail fits in educational communications. Direct mail is a personal contact. It's the next best thing to a personal visit or phone call. With direct mail, you can make personal contact with a far larger number of people than you can through visits or phone calls. Suppose you have a message for Grade A dairy farmers in your state and want to make sure they get it. You can't possibly visit or phone all of them. So the next best way to make sure they get your message is to mail it directly to them.

If you want your readers to take action, direct mail is one of the best media you can use. Direct mail is an action medium. You can use it to get people to try out an idea. Industry uses direct mail to sell products and services. You can also use it to "sell" ideas.

The biggest advantage of direct mail is that you can single out your audience and send it your message in the form you choose. You decide who receives the message, what the message contains, how it is presented, and when it is sent.

Fig. 5-1. Direct mail is a personal contact. In terms of audience impact, it's the next best thing to a personal visit or phone call.

You can come closer to getting desired action with direct mail than you can with any other medium. If you send your message via the newspaper, the editor may not use the story, or he/she may bury it on a page where your intended readers won't see it. With radio or TV, any number of your intended audience may not be listening or watching.

You can be sure of reaching your audience if you select the audience you want to reach. Many people say that audience selection and the mailing list are the most important ingredients of direct mail. And there's little room for doubt about this. You can have a good idea and send out a beautiful four-color mailing. But you won't put your idea across if you don't send it to the right people.

The next important ingredient of direct mail is a good idea. A good idea is one that your readers need, want, or can use at the time you are writing to them. It's possible that a need can be created by playing up a want or some unfulfilled wish. You may have to create the want. You have to show a benefit to your readers—in terms of more income, more leisure time, or more of something else that appeals to them.

Another important ingredient of direct mail is a good approach. A good approach means that your message is written in the readers' language—in terms they understand—and that it stresses benefits to them. Your message should have an easy-on-the-eye layout. Perhaps you need a simple illustration or a big headline to catch their attention or to drive home a point. There are hundreds of ways you can dress up your message, at little or no cost.

HOW TO IMPROVE YOUR DIRECT MAIL

1. Know exactly what you want your mailing to do. What are you trying to accomplish? Are you alerting your readers to some new research development which may interest them? Are you trying to get them to buy a new product?

   It's surprising how many mailing pieces keep their objectives hidden. These mailings often are prepared by people who know what they're trying to do—but who keep what the recipients are supposed to do after they have read the message a deep, dark secret.

   If you want your readers to do several different things, use a continuing series of letters. Sell them one idea at a time.

2. Let your readers know what your idea will do for them. Appeal to the things they are interested in. Give them all the information they need in order to take the action you want them to take.

   A direct mail specialist for a lawn seed company has written in one short phrase a complete definition of copy that sells. He says, "In our copy we must never forget for an instant that people are interested in *their* lawns, not in *our* seed."

3. Make the layout and format of your mailing tie in with your overall plan and objective. Have you used black-and-white when color printing would have been more useful? Have you used color when black-and-white would have done as well? Have you used mimeographing when you should have used another method?

4. Address each mailing piece (correctly) to those persons who can use your idea. Obviously, your first job here is to compile a complete, accurate mailing list, with names spelled correctly. After you have the list, you have to keep it up-to-date.

5. Make it easy for your readers to take action. If you want them to send for a publication, give them all the details of where to send and what to ask for. Take all the guesswork out of the letter for your readers. Put yourself in the readers' shoes and ask, "Now what do these people want me to do about this?" Then make it easy for them to take action.

6. Tell your story over again. Few salespersons make a sale on their first call. Even the best of them call back many times before turning a prospect into a customer. And it isn't reasonable to expect one mailing to produce a large return. If your objective is sound and your mailing list is good, you'll get more results from a planned series of mailings than from a single shot.

7. Research your direct mail. Is your direct mail reaching the right people? Does it contain ideas they need or can use? Are your messages being read, understood, and acted upon?

◄ The Mailing List ►

You can't expect people to buy your product or idea unless your direct mail reaches them. Therefore, an accurate and up-to-date mailing list is the best friend a direct mailer ever had.

Mailing lists contain names of real people. People move, sell businesses, farms, and homes. They buy new ones, go out of business, retire, and die. Merchants drop old items, add new ones; individuals may shift interests to learn about new subjects, new products.

If you want your direct mail to hit center target, you must keep your mailing list as up-to-date as this morning's newspaper.

A simple way to check your mailing list for "deadwood" is to send a questionnaire or card to each person on your mailing list. Ask if he/she wishes to continue receiving your direct mail pieces. Do this once a year.

You may also secure information on removals and corrections from your local post office. Ask your

Fig. 5-2. Throw out the "deadwood" annually. An accurate and up-to-date mailing list is the best friend a direct mailer ever had.

postmaster about Postal Form 3547 and how it can help you to keep your mailing list active.

Be sure to add zip code numbers to your list.

BUILDING A NEW LIST

It's likely that you already have a mailing list in your office. But if you're starting from scratch or wish to enlarge your direct mail business, here are some ideas for you.

The best place for names is your own office letter files. These will give you names and addresses of people interested in your products and your organization. Also, check local stores selling specialized goods to the people you want to reach. Your local library has state and national directories listing names of people active in certain businesses and organizations.

If you wish to use direct mail as a part of your educational program, make sure that names of radio and TV stations, daily and weekly editors, magazine editors, and other media outlets are on your list.

Use names of your local dealers, leaders, and co-operators. Contact these same people for names of customers, neighbors, and friends who might be added to your lists. Check membership rolls of service clubs; business groups; farm, home, and youth organizations. Try county clerks for lists from tax records, birth registrations, marriage licenses, building records, and the like.

Keep names of people who write to you for educational or promotional publications. The publications requested will tip you off as to the individuals' inter-

ests so you can put their names on the right mailing list.

Many businesses use giveaway publications and "sign-up" prizes at public gatherings just to collect these names for their promotional mailing lists.

Don't forget the telephone directory. A rundown of the yellow pages will remind you of businesspersons who should be added to your lists. Do a quick run through the regular section of the telephone book too.

ONE LIST OR MANY?

Some people favor having one grand list and mailing everything to everyone. Others divide their lists into interest groups and select the groups to receive a certain newsletter, card, or promotional piece.

We believe the interest groups are best. They prevent you from flooding your audience with all the pieces of direct mail you produce. They save you time and money because you tailor make your mail to one well-identified audience. And since you are sending out mail pinpointed at an interested audience, you gain in readership.

USING A CODE

Remember to code your mailing lists so that when you must remove names from them, you'll know exactly where to look. It's also a good idea to have some indication on each address plate as to when you added the name. Thus, *HC 80* might mean "homemaker interested in clothing—added in 1980." *GVT 81* might mean "Vermont garage owner, added in 1981."

KNOWING YOUR REGULATIONS

Make certain that you keep up with new postal regulations. This knowledge is vital to anyone using direct mail. If you are a government employee permitted to use "penalty" mail, it's most important to know what you can and can't do with this type of mail. If you pay postage out of your own pocket or that of your organization, you can save money by keeping up on new laws and new rates.

The easiest way to do this is to check with your local postmaster while your mailing piece is in the planning stage. He/she can tell you if your proposed direct mail piece will conform to bulk mailing regulations and can advise you on the most economical way of mailing it. Don't wait until the piece is ready to mail. Check with the postmaster as you plan it.

YOUR STYLE SHOWS

Your word choice and the way you express yourself when you write is termed "style." You should develop a style based on two factors. First, use words and terms familiar to and popular with your readers. Second, say things in your own personal way. Above all, don't mimic someone else's style.

Develop a practical, not literary, style. The manner in which you put words together should result in a clear and readable message. Here are some tips.

Don't take your readers for granted. Things are clear to them only if they understand them. Never pass over a complicated sentence just because you (the writer) know what you mean.

Be sure your readers get the meaning you intend. If in doubt as to its clarity, try the sentence or idea on a friend who doesn't have your technical knowledge of the material.

Even simple things sometimes have to be made clear. Use specific words that have exact meaning for your readers. Although your readers' vocabulary may be as good as, or better than, yours, it may also be different.

Use words you are sure they would use or at least understand. Then your message will be clear to them.

If your readers understand what you're saying without having to reread or study phrases, you're more apt to keep them with you. If they can't follow your writing, you can be sure they'll "put it aside to look at later on." And it will never be read.

Don't go the long way around. Use active voice. *Write:* "Plant the seeds in rows," *not* "The seed should be planted in rows."

The more direct you are, the easier it will be for your readers to follow your thoughts and grasp the message. "Talk" to specific readers. Use words and references with which they are familiar.

GET IN STEP

How many times in the past few months have you started a direct mail letter with these uninspiring words: "A meeting on _____ will be held at _____ in _____"? Ever figure out how many people read no more of your letter than that. What did you do wrong? We'll tell you—**you never got in step with your reader.**

When writing direct mail, try to forget your problems and think of **your readers.** Put yourself in their shoes—and keep **them** in mind throughout your letter.

Use an opening paragraph that promises to **benefit** your readers: Save them time, trouble, and/or money; help their families; solve their problems; etc.

TWO HURDLES

Two major quirks in human nature that are obstacles to your direct mail letters and cards are human inertia and mental befuddlement.

Human inertia. This means pure laziness. People **are** lazy. They don't want to do anything they're not interested in at the moment.

They don't want to read what's in your "envelope." They'd rather go fishing, play golf, gossip, eat, or watch TV.

If more of us realized this, we would send out fewer dull, tiresome, boring letters.

As a letter writer, you've got to *stimulate* your readers to **jump** out of their boredom.

Fig. 5-3. Get in step with your readers, and at the same time overcome the problems of human inertia and mental befuddlement. Stimulate your readers to "jump out" of their boredom.

Mental befuddlement. This is caused by surrounding competition. Today, many sources are competing for the readers' attention. Think of the papers, magazines, letters, and other mail that are put into your own mailbox at the same time. What would be your reading order? And how far down the list would your last direct mail letter be driven?

Let's say that your letter was lucky enough to arrive in the mailbox alone. You're still in trouble! There's competition for attention from the kids, radio, TV, daily paper, dog in the flower garden, and door-to-door salespersons.

You can't afford to put out dull or ordinary letters if you want to educate people by mail. To overcome the readers' laziness and mental befuddlement, you must design and write your letters and cards to get attention and keep it. This takes much patience, lots of trying, and a great deal of imagination.

HAVE A PLAN

If you want a direct mail letter to get results, it'll take some hard work and planning. Effective letters don't just happen. They are designed, planned, written, rewritten, revised, and worried over.

As with any other piece of writing, it's best to set down the goal of your letter first. Write down this objective and how you plan to reach it.

Then take a long, hard look at your audience. Is it composed of parents, leaders, teenagers, retirees, businesspersons, homemakers? You'll need to know all you can about your readers so you can write in their language and appeal to their needs.

Think about how you can aim your letter so it will show a benefit to this audience. This will be the starting point of your letter, so pick the best benefit you have to offer. This could be a free publication, an opportunity to "test" your product free of charge, a financial savings, a way to become more popular, or a method of saving time in work. Always write in terms of your readers and their benefits, not yours.

Remember that people are motivated by certain "drives." Among these are the urges to have financial security, good health, leisure time, popularity, good looks, praise from others, and business and social success.

Notice how the professionals go about pointing out reader benefits:

> Here are two tried and true methods of making money at home.

> Would you like to trade your worries for greater security and happiness?

We are offering you an exclusive free gift in order to introduce our new line.

Open an . . . account and enjoy a new kind of credit shopping. Accept this invitation and get a copy of our catalogue absolutely **free**.

Here's how to put extra money in your wallet. . . .

It's *your* money that's involved, and the stakes are **high!**

While you are thinking about benefits to your readers, also consider some of their objections to the product or idea you are trying to "sell" them. Figure out how you can counter these objections in your letter and build their confidence in you and your organization.

MAKE OPENING VITAL

Remember that, except when you are writing to someone who knows you, the opening sentence is vital. Some direct mail specialists say that in importance the opening carries 90 percent of the weight of the letter.

Fig. 5-4. The opening is vital. Some direct mail specialists say that in importance, the opening carries 90 percent of the weight of the letter.

Don't be discouraged if an opening doesn't come easily. Write a half dozen or more; then pick the best. Keep this opening short, no more than two or three sentences.

Besides offering a benefit, two other effective ways of writing an opening are:

1. Ask your readers a question in such a way that they will mentally agree with you.

2. Include a recent news happening in your lead paragraph and tie it in to your readers' interests.

After you've written that important first paragraph, test it on yourself by asking, "Is this what I'd say to a person I'd just met?"

GET TO THE POINT

Come to the point of your letter quickly. Don't waste your readers' time with a story or attempted humor.

Be enthusiastic and sincere in your writing. If you don't show your own enthusiasm for the product or idea you're talking about, you can't expect your readers to. That enthusiasm must be natural. You'll build confidence if you don't exaggerate, overurge, or oversell. Good letters, like advertising, must be honest, not tricky.

THE HURRY ACTION

If you want your readers to take action as a result of your letter, you must show them some good reason why they should act today and not next week. This can be done by setting a deadline, indicating a limited supply (if there is), or explaining the advantages of quick action.

Do you want them to return a post card or a coupon, ask for a publication, telephone you, send you a check, attend a meeting? If so, carefully explain why that action is important. And make it easy for them to take that action.

Study the experts. Each day you receive several pieces of direct mail. Keep the best for ideas on layout, writing, action, and appeal.

◀ Layout for Direct Mail Letters ▶

Now that you have planned your direct mail letter, the next step is to figure out how you'll catch your readers' attention. You can't sell your idea if the letter isn't read. And unless you can get their attention, it won't be. Competition at the mailbox is terrific these days, and you'll be fighting letters from big business for attention.

This is where a good layout can help you. With very few exceptions, it's the written part of your letter that should take top place in your layout. Layout should be arranged to complement the message and flag down the readers' eyes.

Too often the layouts of amateur letters are too cluttered and complicated. Keep your layout design simple. Complicated design steals attention from the written word.

Many times a lettered heading is all that a letter needs to get attention. A drawing or a photo, if well done and in tune with your message, can also catch the readers' eye and lead them into your letter.

CLIP ART

We have found that clip art books are one of the best sources of drawings for circular letters. This art is done by talented artists and is well worth its small cost. The addresses of firms selling clip art can be found in copies of the magazine *The Reporter of Direct Mail Advertising*.

The Agricultural Communicators in Education has prepared a book entitled *Clip Art Book Number 5*, which can be ordered from The Interstate Printers & Publishers, Inc., Jackson at Van Buren, Danville, Illinois 61832-0594.

Remember that the big reason for a layout isn't to win an art contest but to grab and hold your readers' attention. A good heading or drawing is interesting to look at and moves the readers' eye down into your printed message. Like your copy, a drawing or heading can hold out a promise of something the readers want—an urge for something.

WHITE SPACE

Watch the professional designers who do makeup for magazines, direct mail, and advertising. Notice how they make strong use of white space. Do the same in your direct mail letters.

We've seen effective letters with no headings, no drawings—only extra wide margins and double and triple spaces between paragraphs.

PREPRINTED MAILERS

When you are looking for ways to add color and zip to your letters without straining your pocketbook, try preprinted mailing pieces. These are "tailor-made" letter-sized paper, envelopes, or post cards with art work and headings already printed. There are many firms that supply these items at a reasonable cost. The drawings and captions are designed by skilled artists and printed in four colors. Usually these preprinted items can be run on any standard office duplicating machine. Again, look in

The Reporter of Direct Mail Advertising for names and addresses.

COLORED PAPER

Color adds appeal to almost any written material. The simplest way to add color to your direct mail letters is to use light-colored paper.

When sending out a series of letters or newsletters, you might print your masthead in one or two colors and on a colored paper.

Like everything else you do in direct mail, the design and the choice of a color or no color will depend on your audience. Make certain that what you select will have the right effect on your readers.

◀ Newsletters ▶

Newsletters have gained an important place in the teaching and promotional programs of many organizations. For example, The California Extension Service issues 380 different newsletters from its farm and home advisors' offices. Many of these go out monthly; others at varying intervals.

The same is true of churches, service clubs, and manufacturers. All find that a regular letter with a newsy approach makes effective communication.

Studies show that not only are newsletters read, but about 25 percent of the recipients file them for future use.

CONTENTS VARY

Contents of newsletters vary with the writers and the subjects. In general, they tell of recent developments in the subject matter field, report research and other findings of interest to the readers, carry success stories of individuals, promote products or events that are upcoming, and relay useful ideas. Others answer frequently asked questions, saving correspondence and conversation, and condense or extract from important legislation, speeches, or current events articles. Usually the subject matter is well diversified.

WRITING PATTERNS DIFFER

Newsletters follow a free, less formal type of writing. Many have a combination of the chatty style of a

personal letter plus the brevity and "get-to-the-point" of a news story.

The writer usually assumes that the readers know something about the subject and can build on that knowledge from one issue to another. New or uncommon terms should be defined quickly or used so that their meanings are clear.

As with most communication methods, you must capture the attention of your readers quickly. One easy way to do this is to place the most important, interesting, or timely items first in the letter. If an article is overly long, consider using it as an enclosure with the newsletter.

WRITING

Start with a brief, to-the-point paragraph. Vary your opening sentences. Keep your sentences and paragraphs short for easy reading and more attractive appearance.

Get to the point. Avoid long introductions. Use familiar words. Speak the language of your readers.

HEADINGS AND ILLUSTRATIONS

Use brief, lively headings. Allow plenty of white space around your headings. Subheads may be needed for a few longer articles. Remember that verbs add life to headings.

Here are a few samples:

DIET FOODS ARE GAINING WEIGHT

. . . in the marketplace. Sale of low calorie food will rise 400 percent, from $365 million to . . .

STUCK WITH STAMPS?

Supermarket News, the food trade journal, reports on a major check of cities where trading stamps . . .

HOW OLD IS A FRESH EGG?

Two years or more, perhaps, say Cornell scientists. No storage limit has yet been discovered for their new high-vacuum packaging process . . .

Sketches, simple charts, and graphs can be used. But only if they add meaning to your story and save words.

The picture story makes a good device to get readers interested. If your newsletter is duplicated by the offset printing process, using pictures doesn't cost too much. Make your photo selection very carefully.

APPEARANCE

Naturally, you'll want a neat, attractive letter. Consider these ideas.

Have an easily remembered title and heading design, with a simple, clean-cut illustration of a product, outline of your state, or other emblem that means something to your readers and to your organization. Use the services of a commercial artist for this. His/her work may cost $50 to $150, but spread out over several years and thousands of readers, this is low indeed. The heading design can make or break the looks of your newsletter.

You can add color by printing your newsletter on colored paper. The difference in cost between this and white paper is very little. Use the same color for each issue. This will help your readers to identify your letter.

If you print or mimeograph on both sides, use a paper with enough weight that there will be no "show through."

Use a two-column format for easy reading. If you must use a one-column layout, have extra wide margins and short paragraphs.

You can break the monotony of the column by occasionally indenting statements or other material you'd like to have stand out.

Make good use of white space by not cramming too much on one page.

BE A GOOD EDITOR

The extra effort you put into writing and editing your paper will mean less effort on the part of your readers. Make clarity the main goal of editing. (See Chapter 2 as an editorial guide to this task.)

Have one or two other straight-talking people read each issue and give you their impressions. The message may seem perfect to you, but your readers may get tangled up. Make sure your reading editors are good critics.

If you are presenting information for the first time, make sure you get an official O.K. from the source of your information.

◀ Business Letters ▶

How much does each hand-typed letter you write cost your organization? Research by a well-known corporation shows that the price tag averages $14. This figure is based on a medium-range salary of the letter writer. The higher the salary, the greater the business letter cost.

Obviously, wordiness and unclear thinking and writing push this cost even higher. Imagine the money that's wasted when a letter can't be easily understood by its reader. Unfortunately, this is true of too many business letters.

"The difficulty is not to write, but to write what you mean; not to affect your reader, but to affect him precisely as you wish."

This quotation by Robert Louis Stevenson sums up why it's important to write good letters. Yet, how many of us can report that our letters say exactly what we want them to say and affect readers exactly as we want to affect them? Chances are the percentage is low.

Frequently, letters are written for two reasons: (1) to ask for information, or (2) to give information. Neither is difficult to do. But sometimes even the simplest task becomes drudgery if we don't enjoy doing it. And we usually don't enjoy doing something if we don't know how to do it well.

Too often when we sit down to write a business letter, we completely change our personality. Instead of being friendly and cheerful, we become cold and formal. We fuss and fret and oh and ah and finally try to imitate a letter we've seen somewhere. For example,

"We beg to advise and wish to state that your letter of recent date has arrived. We have noted its contents and herewith enclose the information you requested."

Business letters shouldn't be filled with terms like "regarding the matter," "due to the fact," "I beg to remain," and so on.

THINK BEFORE WRITING

Just as with any kind of communication message, we must analyze what we want to say before sitting down to write a letter. We might ask ourselves these questions: (1) What am I trying to accomplish in this letter? (2) How can I best accomplish it?

There may be other questions we can ask ourselves; but once we have the answers to those questions firmly in mind, at least we're closer to being ready to write than we otherwise would be.

For example, a letter should have an opening that belongs to the readers. The *middle* belongs to the message. And the *ending* belongs to the writer.

The *opening* belongs to the readers because it plays the same role as the lead in a news story. It must catch the readers' attention, indicate what the letter is about, and set up a friendly and courteous tone for the whole letter. And it should link up with previous correspondence by mentioning the date or subject.

The opening paragraph should be short, and it should say something. Here are some examples:

"Thank you for your request for information about silage."

"Here is the bulletin you asked us to send."

"Congratulations on the fine progress your annual report shows."

"I would certainly appreciate having a copy of your report about the new program."

The middle of the letter contains the "meat" of your message. It elaborates on the opening paragraph, as shown in the abbreviated example which follows.

Dear Mr. Jones:

We are happy to send you the booklet you requested.

As a hog producer, you will be especially interested in page 14. It summarizes a report of the university's experiences in raising disease-free hogs.

Won't you write us again if you would like more information?

STOP WHEN MESSAGE IS COMPLETED

Many letter writers repeat the same ideas in different words. They are as annoying as the visitor who says, "Well, I've got to leave . . ." and then sits and talks for another hour.

So, you should stop when you've completed your message.

Fig. 5-5. Stop when you have completed your message.

Direct statements with active verbs are the most effective endings we can use. Participle endings are the least effective. They are weak, hackneyed, and incomplete in thought. Two examples are: "Thanking you in advance" and "Trusting we shall have your cooperation in this matter, I remain . . ." Such endings are as bad as opening statements that say, "We have here before us your letter of recent date" or "As per your recent request . . ."

Here are a few examples of direct, effective closings.

"If we can help you in any way, won't you let us know?"

By using a question, we *invite* the receiver to write

again. By saying ". . . please let us know," we merely give the receiver permission to write again.

"We will appreciate knowing by June 6 so that we can plan the remainder of the program."

Before using such an approach, however, we might consider whether we want to open up opportunities for further correspondence. A pen pal may be the last thing we need.

PITFALLS TO AVOID

There are also other pitfalls that we should avoid in order to become better letter writers. These include using jargon; writing letters by hand; using poor grammar, spelling, and punctuation; and delaying answers to mail.

Jargon. This consists of stilted, overused words or phrases, or terms that are familiar only to us or to our organization and that are misunderstood by others. Jargon makes reading stiff, formal, and dull.

Some examples are:

"According to our records . . ."

Instead, say, "We find . . ."

"Enclosed please find check."

This implies that the reader must hunt for the check. "We are enclosing our check" is sufficient.

We can avoid trite expressions by writing as if we are talking to the person. In talking, we wouldn't say, "Thanking you in advance for the circular . . ." We'd more likely say, "I'd appreciate receiving a copy. . . ."

Handwritten letters. None of us will ever regret learning how to use a typewriter, for not everyone has a secretary. If we must handwrite our letters and notes, we should do it neatly.

Poor grammar, spelling, and punctuation. These reflect unfavorably on the writer's intelligence and ability to think clearly.

Delayed answers. An excellent policy is to answer first class mail within three days.

Photo / Art Credits

Bill Ballard, North Carolina State University: Figs. 5-1, 5-2, 5-3, 5-4, 5-5.

SPEAKING

◀ Correcting Speech Faults ▶

Few people speak as well as they could. Do you? Have you ever said to someone something you thought was perfectly clear, only to discover later that you were completely misunderstood? Sure, it's happened to all of us. But what was your analysis? Did you say the misunderstanding was the listener's fault—the listener just wasn't paying attention? That may be true, but the fact remains that you failed to communicate. The burden is largely on you to assure understanding.

Physically, there probably is nothing wrong with your speech mechanism. Few people today have physical speech defects. The problem more likely is how you're using what you were born with.

It's an unusual person whose speech can't be improved. Poor speech is learned, just as is good speech. If you practice poor speech, it becomes a habit. If you practice good speech, it too will become a habit. That's why it's so important to become speech conscious. This is the first step toward better, more intelligible speech.

Listen to yourself. Tune your ears to your tongue; really listen to what you say. You'll be surprised at some of the things you hear. You only have to listen to a tape recording of yourself to know there can be and often is a difference between what is thought and what is said or heard.

Remember that just because you speak, people don't necessarily hear and understand.

THE PROBLEM: SIMPLE WORDS

The most powerful words in the English language are the short ones—"live," "love," "hate," "die," "war," etc. And surprisingly, it's the short words which are most often misunderstood. Why? Short words, more than long ones, have "sound-alikes."

Test the principle of "sound-alikes" for yourself. Read aloud the following word groups: "forge, forage, 4-H, forehead"; "shell, shelf, self"; "leave, leaf, left"; "cook, crook, brook, book." Say them again; this time hold your fingertips lightly to your lips and face. Feel the different lip, mouth, and jaw movements and the resulting puffs of air. In addition to hearing the difference, you can feel it. Do the test in front of a mirror and you can see the difference too. Each word has a different feel and speech mechanism requirement.

Poor intelligibility occurs when some part of the word is sounded out incorrectly, incompletely, or not at all. This is the cause of miscommunications.

To help you become more aware of your speech mechanism and what it is or is not doing well for you, read rapidly and loudly the following word sets. Listen and look at yourself carefully. Is each word clearly distinguishable? Are the mouth parts making the necessary movements? Listen as you look and feel with your fingers. Be especially sensitive. Notice the differences:

Prepared by **J. Cordell Hatch,** electronic communications specialist and professor of agricultural communications, The Pennsylvania State University.

cave	bridal	pure	quarter
cake	final	poor	porter
cage	vital	tour	water
case	title	two	order
tack	yes	cracker	left
tact	jet	chatter	list
pact	get	shatter	lisp
pack	yet	shadow	lid

The reading of word groups like this, the "how now brown cow"–type sayings, or words listed in a thesaurus or dictionary, will help to sharpen articulation and make you more speech conscious.

RATE YOURSELF

After you've practiced reading, speaking, and listening for awhile, rate yourself. You may want to record your practice sessions. Listening to the tapes will give you a good feel for your strengths and weaknesses. If possible, videotape and critique your practice sessions.

Joe Tonkin, audio-visual specialist with the USDA Extension Service for many years and a major contributor to speech and broadcast training, said, "Speech is a personal, almost magical thing bound up with our private thoughts and feelings and ideas. To study and improve speech we must take a detached, impersonal attitude toward it."

Any efforts, he said, to improve our speech must begin with an honest and critical self-appraisal. What are your speech faults? The following are several common ones. How many do you have? How serious are they? Check yourself.

Do you:

( ) slur words?

( ) mumble?

( ) grunt sounds?

( ) swallow sounds?

( ) whisper rather than speak out?

( ) talk too fast?

( ) talk too slowly?

( ) talk in a monotone?

( ) have lazy lips?

( ) talk with your lips closed?

( ) have noisy lips?

( ) speak through your teeth?

( ) articulate poorly?

( ) need to speak more forcefully?

( ) let your breath out too quickly?

( ) sound unenthusiastic?

( ) speak in a halting, hesitating manner?

( ) give the impression of uncertainty?

Again, a tape recorder is useful in analyzing your speech habits. A videotape recorder is even better. With it you can see face and mouth movements at the same time you hear the words. Participate in toastmasters and speech clubs, as well as in courses, if you have the chance. The following units should be helpful also.

PHONETICS AND INTELLIGIBILITY

Phonetics is the science of speech sounds. There are two kinds of phonetics: physiological and acoustical.

Physiological phonetics is concerned with speech energy and the way we make speech sounds. The source of speech energy is the lungs. The stomach muscles force the air from the lungs upward to the vocal cords, thus allowing the stomach muscles pressing on the lungs to control loudness. Vocal cords control the size of puffs of air that enter the vocal tract. The faster the frequency of these puffs of air, the higher the pitch will be.

From the vocal cords, the sound moves into the resonant tract of the throat. It's in the resonance that we find the first indication of individuality in speech.

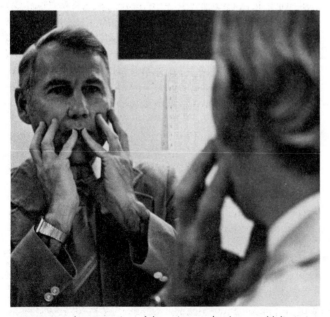

Fig. 6-1. A demonstration of the voice mechanism sensitivity test.

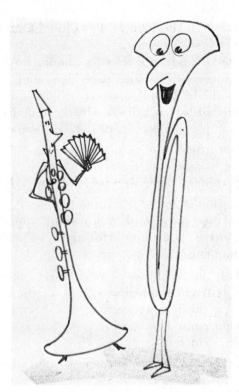

Fig. 6-2. One person is a clarinet; another is a trombone.

word "notice," the "t" often sounds like a "d." In words like "hunting," the "t" often degenerates into a grunt.

Many times we are misunderstood because we don't say the whole word. We don't pronounce the first or last letter. With many similar consonant sounds, such as "t," "d," "c," "b," "g," etc., it's obvious that there's an infinite possibility of phonetic error. This is also true in the confusion of "g" and "w" sounds. Here we're dealing with a problem of articulation.

Fig. 6-3. There are four types of articulation: plosives, fricatives, laterals, and vowels.

Like a fingerprint, the voice resonance of each person is unique. One person is a clarinet. Another is a trombone. We can't substantially change this resonance. We can only make the best possible use of it.

Finally, sound becomes speech as it passes the articulators—the tongue, the lips, the palate, the teeth. Here the forming of words takes place.

There are four types of articulation in English: plosives, fricatives, laterals, and vowels.

1. *Plosives.* The breath stream is completely stopped by closing the passage. The "p" in "pie" and the "g" in "go" are examples.

2. *Fricatives.* A narrow slit or groove is formed for the air to pass through. Examples are the "th" in "thin" and the "sh" in "shin."

3. *Laterals.* The middle line of the mouth is stopped, but an air passage is left around one or both sides. The "l" in "let" is an example of a lateral.

4. *Vowels.* The passage is left relatively unobstructed. Vowels are usually contrasted with consonants. Plosives, fricatives, and laterals are all consonants. Vowel sounds determine the quality of our speech. Vowels are the sounds of maximum resonance.

Unintelligibility is generally caused by not articulating consonant sounds. For example, in the

Acoustical phonetics concerns itself with the speech sound after it has left the mouth. The strength of this sound is measured in decibels. Ordinary conversation at a distance of 3 feet is about 60 decibels of sound.

Sometimes it's a good idea on radio and television to use a little more force in your speech—about 10 decibels. Awareness of loudness, pitch, and rate can improve listenability as well as intelligibility. Also, when speaking before groups, try to get a feeling of "side tones," so you can make adjustments in your voice appropriate for the size of the room, size of the audience, and acoustics of the room.

PRACTICE FOR MORE INTELLIGIBLE SPEECH

The context within which words are used is sometimes sufficient to give meaning to poorly pronounced words. The wise communicator/educator—good speaker—avoids the risk by developing intelligible speech and by seeking constant improvement.

The following points are worth repeating and remembering. They should be practiced in everyday as well as formal speech. Intelligible speech in interpersonal communications is just as important as when you are addressing a group or speaking on the radio or television. Only through practice will good habits become internalized.

- We learn monosyllable words earlier and use them more often. Since they have more "sound-alikes" and alternate meanings, they are the words most often mispronounced and misunderstood. Polysyllabic words, on the other hand, have more acoustical cues for discriminating between words. Concentrate on the short words!

- Sound is energy; it passes through the air in different wave lengths. The faster the puffs, the shorter the wave lengths and the higher the pitch. Pitch and volume may be varied; it's speech variability which adds interest and understandability to what's said. A soft, monotone speech often gives the impression of speaker uncertainty. Vary pitch, loudness, and speed!

- Poor speech is sometimes caused by the speaker's inability to hear well. Acoustical feedback from the inner and especially outer ear allows for modifying the voice and making necessary adjustments. A keen awareness of these "side tones" is essential for good speech. This is why a radio or TV announcer sometimes wears a headset or cups a hand between the mouth and ear to catch more of what is being said. Listen to yourself!

- The pacing or pauses in speech are important. They can add dramatic impact and understandability. However, they border very closely to halting, uncertain delivery and can reduce speaker credibility and confidence. Fast talkers, about 180 words per minute (average is 150), are more highly regarded, more persuasive, preferred, and obviously can communicate more in less time. With speech usually comes enthusiasm. What's your speed? You may want to pick up the pace a bit.

- Poor readers tend to be uniform; trained readers show variability. The more adept you become at expressing thought units, rather than mere words or lines, the more interesting and distinct you will be. Vary the sound and your emotions to suit the meaning of the words being read or said. Loudness, pace, and word stress reveal attitude as well as prevent misunderstanding. Example: "I **favor** the idea!"

- Use abdominal muscles rather than the diaphragm and upper chest to provide a steady breath stream and the power needed for good speech.

- Throat and neck muscles should be relaxed. However, the mouth parts (lips, tongue, jaws) should be mobile and active in shaping each puff of air or syllable clearly and distinctly. Examples: "ro**b**e," "ro**p**e," "**t**ake," "**b**ake."

- Poor articulation results when your "t" sounds like a light "d." Examples: "no**t**ice," "ra**tt**le," "Sa**t**urday," "mor**t**al," "migh**t**y," "hi**t**," "i**t**."

- A grunt is a poor substitute for the "t" in these words: "impor**t**ant," "hun**t**ing," "quan**t**ity," "den**t**ist," "ki**tt**en," "foun**t**ain," "plan**t**ing," "moun**t**ain," "twen**t**y."

- With the retroflex "r" as in "ca**rr**y," "te**rr**ible," or "cu**rr**ent," it helps to think of the word as being divided between the "r's." Try it with a slight pause between the "r's": "car **r**y," "ter **r**ible," "cur **r**ent."

- And those triple consonants—don't let them throw you, as in: "wasps," "wisps," "lisps," "insists," "resists," "consists." These are mostly tongue-and-teeth words.

Finally, avoid Slurvian self-taught.

WHATCHADOIN?	What are you doing?
NOTHINMUCH.	Nothing much.
WHATCHAGOINADO?	What are you going to do?
GOINFISHIN.	Going fishing.
THEYKETCHINENNY?	Are they catching any?
DONNO.	I don't know.
JEET YET?	Did you eat yet?
NO, JEW?	No, did you?
OKAY, GIMMINEG.	Okay, give me an egg.

SUMMARY

Our main concern is with "intelligibility." We're not dealing with dialect, because if dialect is properly enunciated, it can still be intelligible.

No physical part of the resonance or enunciation tract controls the warmth of our voice. That's in our mind—our attitude—our feeling toward our listeners.

When we read, we sometimes read in a monotone. This is caused by reading one word at a time. Try to see clusters of words as thoughts, then sound them out with the appropriate emphasis.

Good breath control for speech requires abdominal, not diaphragm, breathing. When we exhale from the diaphragm, we heave, or pant, and our voice becomes jerky.

Although we can't make changes in resonance, we can refrain from mumbling, lip laziness, and lack of enunciation—habits that ruin voice quality.

If someone says you sound breathy, speak more loudly. If your voice sounds thin, open your mouth and lower the pitch when you speak. If your voice sounds harsh, don't lower your pitch and breath support at the end of sentences.

Here are five guideposts to better speech:

1. Learn what to listen for in speech—acoustical qualities and variables.
2. Listen critically to other people to develop your own awareness of speech characteristics.
3. Listen critically to yourself to become aware of your own speech characteristics.
4. Experiment with your own vocal output.
5. Repeat improvements in your own speech until they become habit.

◀ Speech Delivery ▶

Very few people are able to get up before an audience without some nervousness and discomfort. A little fear and tension are normal, but too much can tie you in knots. To the extent possible, your voice, expressions, and posture should be warm, friendly, relaxed.

Nervousness can be reduced by thorough knowledge and preparation of your topic; a positive attitude toward yourself, your topic, and your audience; and practice. Occasional failure and disappointment are inevitable—they're part of the learning process. But with experience, you'll soon gain the self-confidence and skill possessed by good speakers.

Effective speaking is the *oral* expression of knowledge. A well-informed person can be a useless bore, while an effective speaker will have something to say, and the skill to say it.

There are two aspects of effective delivery: *voice usage*—that which people hear, and *physical behavior*—that which people see.

Through visual impressions, your audience will develop an attitude toward your sincerity, friendliness, and energy. They read this from your facial expression, the way you stand or walk, and what you do with your arms and hands.

Your mental attitude is important also. Think of your audience as you speak to them. Know the content of your presentation well enough to avoid using mental energy remembering what you want to say next. Be interested in the people you're talking to, and keep thinking of them as you speak.

Many speakers make complete preparations in advance, including impromptu remarks. However, reading directly from copy all the time can be distracting to the audience. In general, it's best to refer to notes from time to time rather than to read verba-

tim. A well-planned outline is important, though, as it should help speaker and audience travel in the same direction. Speaking is an art; preparation is a science.

Delivery may be broken down further into voice usage, body action, eye contact, and facial expression.

VOICE USAGE

1. *Rate or time*

   Variety is needed. Many persons use about the same rate regardless of the material involved; rate should fit the mood of the material.

   Pauses are the stop signs, the punctuation marks of speech. Pause to let ideas sink in—for meaning, emphasis. However, when you do pause, do so in silence. Avoid vocalized pauses ("uh's," "er's," "ah's," etc.). Vocalized pauses only call attention to the fact that you're trying to think of what you're going to say next.

2. *Pitch*

   A good speaking voice has range and flexibility of pitch. Pitch your voice so that you can easily raise and lower it for emphasis.

3. *Force or loudness*

   Your voice must be loud enough to be *heard*, but it should also be used as a tool for emphasis. Force and loudness should fit the mood of the material and the occasion.

4. *Articulation*

   Articulation means distinctness of utterance.

Lazy lips, tongue, and teeth cause poor articulation. Don't mumble, mutter, run words together, or go back over words. (See the earlier part of this chapter for other advice.)

BODY ACTION

Body action (posture, gestures, and movement), plus vocal expression, plus knowledge, equals communication.

Make your body work for you in speaking. The act of moving from one place to another and the act of gesticulating with the hands or other parts of the body can direct or attract attention. The eye instinctively follows a moving object and will focus upon it. We watch a moving picture with greater involvement than we watch a still one.

A sleepy audience can be awakened by movement. Movements should be natural and easy. Be careful not to make them a distraction.

Forward or backward movements are often associated with the importance of a point.

A step forward—more important

A step backward—relax; think about that last idea

Movement helps relax you, but try to avoid random movements. There's a difference between fidgeting and meaningful motion.

Gestures and other movements can help reduce nervousness as well as reinforce what you say. A few moments of vigorous exercise, even a couple of deep breaths immediately before your presentation, will help relax you and reduce nervousness.

1. *Posture*

   Posture should never draw attention from what you're saying to how you're standing. Posture which is relaxed, but not slumped—erect, but not stiff—is best. Be comfortable without being sloppy.

2. *Gestures*

   Movements of any part of the body can reinforce your oral presentation. Know the difference between meaningful gestures and a case of the fidgets. Your audience will recognize the differences.

   Gestures help convey meaning, and they increase your own energy and self-confidence. They help hold attention. Descriptive gestures are imitative. With them you describe the size, shape, or motion of an object through imitation. Gestures of the head and shoulder are the most common, aside from universal hand gestures. You may shake your head or shrug your shoulders, as if to agree or disagree with a point.

   What can you do with your hands? Leave them at your side? Put them in your pockets? Clasp them behind your back? Clasp the sides of the lectern? In none of these cases are your hands working for you.

   Use gestures to reinforce your ideas. Make gestures strong. Avoid abortive gestures, ones that never quite get off the ground. Avoid weak, wishy-washy, apologetic gestures. A relaxed body will help to prevent stiff or awkward gestures. Make positive gestures, but vary their intensity to suit the point. A gesture should come with, or slightly before, the words it's intended to support. Never gesture after you have said the word or phrase.

3. *Movement of the body as a whole*

   Movement can be used to indicate transitions between thoughts; but avoid the undesirable types, such as shuffling feet, shifting weight, or fidgeting with the microphone or other equipment. Movement and gesture can be combined in explaining visual aids. In fact, they themselves are visual aids.

EYE CONTACT AND FACIAL EXPRESSION

With your eyes you can attract your listeners. Look *at* and talk *to* your audience. It makes your listeners feel you're interested in sharing information as well as feelings with them, as indeed you should be.

Be sure that your *facial expression,* as well as your voice, reflects your own interest in your speech and in sharing it with your audience. Avoid the "poker face" approach to public speaking.

Because your face reflects your attitude, you can use it effectively to emphasize or support your message.

Concentration tends to bring a frown to the face. Know your presentation well enough to be able to avoid concentrating too heavily on it. Or learn to concentrate with a pleasant expression. Take time to get comfortable before speaking . . . and don't suddenly relax at the conclusion of your talk. This indicates to your audience . . . "Thank heavens it's over!"

Ralph Waldo Emerson once said, "What you *are* thunders so loudly I can't hear what you *say.*" Be sure the things you do with your body as well as your voice help rather than hinder in getting your message across.

When called upon to introduce a speaker, what do you do? First, remember that you are introducing the person, not giving the speech. Introductions have a tendency to be too long and not relevant to the occasion. Everything done and said should contribute to the speaker's effectiveness.

Your objective is to introduce the speaker as a person who has something important to say, who can say it with authority, and who has the ability to inspire confidence in the audience. Sell the speaker and the topic. Stimulate a desire to hear what will be said. Establish the basis for speaker-audience rapport. The introduction doesn't have to be long to do this; it simply turns the spotlight of audience attention on the speaker. Don't bungle the job by saying too much or too little or by spotlighting yourself.

Fig. 6-4. When you introduce a speaker, tell as much about the speaker as is needed, but don't say too much or spotlight yourself.

OTHER GUIDELINES

You can create interest in the speaker and the subject by answering questions the audience is likely to be asking: Who is the speaker? Where is he/she from? How qualified is the person to speak on this subject? Why should I listen?

Who is the speaker? If the speaker isn't familiar to the audience, give his/her name twice—once at the opening of the introduction and again at the end. Be sure to pronounce it correctly—if you're not sure of the pronunciation, check with the speaker before the meeting. The audience will also be interested in his/her occupation, title, or position.

Where is the speaker from? Listeners want to know where the speaker came from originally and where he/she comes from now. These may seem insignificant, but audiences like to hear them.

How is this person qualified to speak on the subject? Point out the speaker's qualifications—experiences, writings, background. Give interesting data, achievements, and the fields worked in, especially those areas relating to the topic.

Why should I listen? Emphasize the importance of the subject. Show a need for information and tie it to the interests of the audience. Bring out personal values by saying, for example: "All of us pay taxes. A knowledge of the way taxes are apportioned is valuable both to our pocketbooks and to our understanding. . . ."

You can end the introduction by formally presenting the speaker to the audience, giving his/her name once again: "May I present a friend of Hale County, James J. Harrell." Previous remarks should have been directed to the audience. Now, and only now, do you turn to recognize the speaker. Remain standing as the speaker comes forward; lead the applause; and then sit down.

WHEN THE SPEAKER FINISHES

You need to add only two or three sentences at the end of the speaker's talk. These should include a brief and sincere thanks, which should be addressed both to the speaker and to the audience. After the meeting, add a few private words of thanks.

Avoid the temptation to give a summary of the talk, take issue with any remarks, or add items. It is all right to underscore briefly an important point by stressing its relevance to the audience.

FORMAL OR INFORMAL

A *formal* introduction has dignity as well as interest. An *informal* introduction relieves tension and sets the tone of the meeting. Whether the introduction is formal or informal depends upon the prestige of the speaker, the occasion, and the extent of your acquaintance with the person. For example, an introduction of someone who is speaking on a serious occasion or of a stranger would be formal. Unfortunately, some introducers know only one type and use the type they know, no matter what the occasion or who the speaker.

In general, the better *known* the speaker, the

briefer your introduction should be. The more *unknown* the speaker, the more you'll have to arouse interest.

Remember—there are four necessary elements in introducing any speaker: *tact, brevity, sincerity, and enthusiasm.*

These suggestions are to help you plan your introduction. Using them will help, but *you* must supply the rest.

WHEN YOU INTRODUCE A SPEAKER, DO—

- Be brief—the audience has come to hear the speaker, not you.

- Use humor only if it suits the occasion, is in good taste, and creates friendship.

- Speak loudly and clearly enough to be heard easily.

- Check the introduction you plan to make with the speaker.

- Be sincerely enthusiastic—but don't overdo it or gush.

- Suit the nature of the introduction to the tone of the speech.

- State the subject of the talk correctly.

- Practice your introduction.

WHEN YOU INTRODUCE A SPEAKER, DON'T—

- Talk about yourself and how you felt the last time you spoke to a group of this size.

- Emphasize what a good speaker or witty person the speaker is—let the performance speak for itself.

- Give your views on the subject of the talk—you'll steal the speaker's material.

- Give committee reports, meeting announcements, etc., with the introduction.

- Apologize because the speaker is a substitute or isn't a known figure.

- Tell embarrassing stories or jokes about the speaker.

- Use trite remarks, such as "Our speaker tonight needs no introduction."

- Embarrass the speaker by an elaborate build-up—you might make living up to your description impossible.

◀ Presentation and Acceptance Speeches ▶

Although you probably present more awards and honors than you receive, it's good to know how to be a gracious receiver as well as a thoughtful giver. Whether you're presenting or accepting, observe all the rules of common courtesy and public speaking.

THE PRESENTATION SPEECH

A well-given presentation speech adds to the honor or award. It publicly expresses the recipient's merit and the donor's appreciation. Therefore, what's said should be appropriate to the occasion.

Name the award. The audience wants to know who the donor is and why the award is being given. If the award is given as a memorial, give all the reasons.

Explain why the recipient is receiving the award. State specific accomplishments. Be accurate and complete, but don't exaggerate. The audience is interested only in the achievements directly related to the award.

Tell how this person's work will influence the accomplishments of others. But avoid comparisons that might embarrass either the receiver or someone else who might have won the award.

If the recipient is an organization, stress the principles it stands for.

Tell why the particular award was selected if it has a special significance to the receiver. Stress the symbolic nature of the award. Avoid mentioning the cost or the difficulty in deciding what to select.

Practice the speech so that you can deliver it well without reading it. Be brief, but complete. Be genial, but not humorous. Be sincere and enthusiastic; if your heart isn't in it, let someone else make the presentation.

Hold your audience in suspense by giving the recipient's name at the end of the speech. Smile when you speak directly to the winner, present the award, shake hands, and repeat the recipient's name.

When you hand the award to the recipient, give it up. Don't make the honoree reach for it.

Fig. 6-5. Spotlight the award recipient, not the presenter.

After you've handed over the award, don't continue to speak. Your speech ended the moment you released the award. If photographs are being taken, you may have to re-enact the presentation. Do this slowly. You and the recipient should face each other but turn slightly toward the camera, stand close together, and hold the pose at the moment of award exchange until all photographs are taken. Then you should move aside so the recipient can have access to the microphone to make an acceptance speech.

THE ACCEPTANCE SPEECH

In an acceptance speech, you must feel and show appreciation. Face the giver as you express thanks; call the giver by name. Also thank the donor, using the person's or the organization's correct full name.

Often a sincere "thank you" is sufficient. But at other times, more may be necessary—especially if you know about the award in advance. Never begin an obviously planned acceptance speech with "I really don't know what to say."

Give credit. Share the honor with others whenever possible. Show how the donor made it possible for you to receive the award. Without giving too much background, pay tribute to associates who helped you. Describe the work of others specifically.

Be modest. Minimize, but don't depreciate, your own merits. Attribute whatever you should to the cooperation of others and to favorable circumstances.

Compare what you have done with what is yet to be done—but don't tell the audience what others should do. Describe a humorous or interesting experience involved in the achievement. But don't embarrass anyone or get so personal that the audience can't enjoy it too.

Make the group feel that you're pleased with the award without referring to its monetary value. Indicate its significance, the responsibility it puts on you, and your determination to live up to the honor. Avoid leaving the impression that it's something you've "always wanted."

Keep your talk brief. Don't lapse into personal conversation or comments with the giver or co-workers.

End your acceptance with a **brief** statement of **sincere** appreciation.

Then sit down, and don't disrupt the proceedings by talking further about the award with persons seated nearby.

◄ Preparing a Speech That Says Something ►

Here are some guidelines for making a good speech:

1. Adapt the presentation to the audience members' interests.
2. Determine in advance the reaction desired from members of the audience.
3. Spend enough time in preparation to make your remarks effective, relevant, and valid.

ADAPT SPEECH TO AUDIENCE

When audience and subject have a real relationship, the speaker can be assured of a favorable hearing. Thus, as a speaker, you need to show how your topic concerns each individual or the entire group.

Material should relate to the purpose of the get-together. If the subject has been assigned, limit remarks to a particular aspect that fits the occasion and the audience.

For example, suppose you've been asked to speak on "quality meat-type hog production." You're to talk to three groups—a farmers' group, a city business club, and a homemakers' organization. Here's how you can adapt the topic to the varying interests of the three groups:

1. To farmers, stress how quality meat production

will benefit them, and explain how to raise quality meat-type hogs.

2. To non-farm businesspersons, tell how better quality meat affects them as consumers and how quality production increases farm income—which may mean better trade for all.

3. To homemakers, explain how farmers provide better pork chops for tables and how to identify quality meat at the supermarket.

Normally you'll be asked to speak on something in which you have some authority. If you're asked to speak on a subject outside your area, you should refuse and perhaps explain why you're not qualified.

IDENTIFY PURPOSE

If you want to convince your audience, the reaction you seek is agreement. Suppose you want to convince farmers of the value of raising quality hogs. To do so, you'd show proof of your statements by logic, reason, accurate and timely examples, recent statistics, and opinions of authorities. You'd also want to demonstrate that your idea is the best available solution. And you'd want to make that idea agree with what people feel is right or decent.

Perhaps you want to stimulate action. Here again, you must convince your audience—but then you must go one step further. The reaction you want is not only agreement, but also a definite, observable action. For instance, not only do you want to convince farmers that they should produce quality hogs, but you also want to persuade them to do so.

If you're planning to inform the audience, the reaction you want is clear understanding.

PREPARE THE TALK

If you haven't time to prepare a talk, don't accept the engagement.

Thorough preparation should extend over two or three days at least. The speech should be entirely completed the day before it is to be given.

Thorough preparation is the best available fear-remover!

Preparation also means analyzing the audience so that you'll be sure to choose the right subject matter, the appropriate supporting material, and the proper approach to your topic. Keep in mind the group's attitudes; they will influence the group's acceptance of your ideas.

There are several ways to get information about an audience. The group's officers and members can give you information, or you may attend one of their

Fig. 6-6. Analyze the audience so you'll be able to choose the proper approach and delivery method for your topic.

meetings. Literature or official publications also can help. Ask others about the group too.

Get attention at the beginning of your speech: Ask a question; make a startling statement of fact or opinion; or use a quotation, a concrete illustration, a humorous incident, or a reference to the occasion or the subject.

The latter is the most common way to begin. For example, say, "Quality meat-type hog production could mean the difference between profit and failure for you."

For the city business club, you could say: "Quality meat-type hog production boosts farm income and generates a larger market for the things you sell to farmers."

A rhetorical question could begin your talk to the homemakers' organization: "Are you interested in serving better pork chops to your family? Of course you are. And farmers are trying to help you get better pork for your table."

Divide your speech into two, three, or four main headings. Most listeners can't remember more than four basic points.

Adjust your language to the audience members' educational level and knowledge of the subject. One way to be sure your subject is not above or below the intelligence of members of the audience is to aim at the average member of the group.

Your conclusion is your last chance to stress the purpose of your talk. You don't want anyone to leave with any doubt about what you've tried to say. So, at the end, focus on the central idea you've developed through your speech.

You might end with a summary, a challenge, an appeal for action, a quotation, a story, an illustration, or a statement of personal intention. Or, you might finish with a question or a problem if you don't have an answer but want the audience to think about the solution.

The conclusion usually isn't the place for humor. But, if you can accomplish the purpose of your speech and still use humor, do so.

Leave the selection of a title until after you've organized your talk. Then make it original, brief, relevant, and provocative.

Practice aloud. Don't practice so much that you get stale, but have material well in mind. Then don't read your speech—talk it.

Instead of working from a script, you may prefer to speak from note cards that list the main points of your talk.

Finally, be succinct. There's nothing worse than a long-winded speaker.

◀ Points for Effective Communications ▶

Successful communications require a high level of attention. Without attention there's no communication. The attention a person pays to a particular communication is a function of three factors.

1. How strongly motivated the listener is.
2. How much decoding effort is involved.
3. How long attention is to be held.

In spoken and written communication, you can reduce effort for the receiver by using shorter words and sentences, more examples and illustrations, and clear and logical expressions. In pictorial communication, you can reduce effort for the viewer by clarity, sharpness, and simplicity of visual images. Make it easy to see at a glance what the message is all about.

New information is easier to grasp if you present it a step at a time, with each point building on what has gone before. Relating unfamiliar concepts to things the listener knows about will reduce effort. Spell out implications. People usually fill in the gaps left by vague or missing information—sometimes correctly, sometimes not. Sometimes they give up.

Redundancy can make a message *easier* to understand. It reduces chance of error and emphasizes the important information. But redundancy can also make a story *harder* if it's monotonous or distracting.

Fig. 6-7. Get attention at the beginning of your speech.

Multiple media presentations—either simultaneous or separate—can provide an acceptable kind of redundancy. For example, a talk with slides, film, acetates, or real objects is better than a talk with no visual aids.

People are more interested in and motivated to learn about matters that are physically and psychologically near to them. People accept a message more readily if they approve of the source and if the message coincides with their preconceptions, prejudices, or prior beliefs.

◀ Some Principles of Persuasion ▶

To accomplish attitude change, you must make a suggestion for change which your listeners will receive and accept. Acceptance of the message is a critical factor in persuasive communication. The suggestion will be more likely accepted if it meets existing personality needs and drives; if it's in harmony with group norms and loyalties; and if the source is perceived as trustworthy and authoritative.

A suggestion made face-to-face and coupled with mass media reinforcement is more likely to be accepted than a suggestion carried by either channel alone, other things being equal.

Change in attitude is more likely to occur if the suggestion is accompanied by other factors underlying belief and attitude. This refers to a changed environment which makes acceptance easier. There probably will be more opinion changes in the desired direction if conclusions are explicitly stated than if the audience is left to draw its own conclusions.

When the audience is friendly, or when only one position will be presented, or when immediate but temporary opinion change is desired, it's more effective to give only one side of the argument. When the audience disagrees, or when it's probable that it will hear the other side from another source, it's more effective to present both sides of the argument. When equally attractive opposing views are presented one after the other, the one presented last will probably be more effective.

Sometimes emotional appeals are more influential, sometimes factual ones are. It depends on the message and the audience. A strong threat is generally less effective than a mild threat in inducing desired opinion change. The desired opinion change may be more measurable some time after exposure to the communication than right after exposure. There is a sleeper effect in communications received from a source some listeners regard as having low credibility. In some tests, time has tended to wash out the distrusted source and leave information behind.

◄ Other Communication Principles ►

People tend to expose themselves to communications which agree with their existing opinions and interests, and they usually avoid unsympathetic material. They perceive what they want to perceive, have habitually perceived, or have been rewarded socially or physically for perceiving. They learn and retain information consistent with their own beliefs and forget what is not more quickly.

Also, research shows the following:

- A "positive" communicator (meaning one well-liked, prestigious, and considered trustworthy) can, at least up to a point, accomplish more change by advocating a greater change.

- Stating conclusions in the message is more effective in changing the opinions of the less intelligent members of the audience than the opinions of the more intelligent ones.

- Explicit refutation, although it makes for clearer understanding of the argument, is also likely to arouse antagonism and may inhibit change. Elaborating on the audience's arguments is counterproductive.

- "Flogging a dead horse" (for example, praising academic freedom to a college faculty before getting to the real purpose of the address, which was to change opinions on governance) helps win over the audience on the major point.

- Persuasive speech containing sound evidence is more effective.

- Recipients of a communication tend to agree with a well-liked speaker, disagree with a disliked one.

- In regard to primacy and recency effects, the earliest serial items and the latest ones tend to be remembered better than the middle ones.

- Arousing needs, then presenting persuasive material relevant to those needs, is more effective than presenting the persuasive material first.

- Placing a message highly desirable to the recipient first, followed by the less desirable messages, results in more change.

- The communicator should give pro arguments first.

- If a person speaks on a topic about which he/she is obviously unqualified, what is said tends to be disregarded.

- Subject matter is the chief determinant of listener interest.

◄ Tips for Good Telephone Communication ►

As the business world becomes more complex, the telephone becomes increasingly important as a channel of communication. When you're really in a hurry, nothing beats this handy, convenient instrument.

USING THE TELEPHONE EFFICIENTLY

Follow these tips in making your telephone time more productive and economical:

1. Because of the live, two-way attributes of the telephone, it's best used where instant interaction is required. It facilitates fast, on-the-spot decision making. Also, it's great for urgent, brief messages that require discussion.

2. Avoid long, detailed, technical communications or complicated figures and data. They can become garbled and misunderstood, and there's no written record for verification or future reference. If possible, arrange for your party to have technical material, lists, and figures beforehand so they can be referred to during the call.

3. On local or long distance calls, resist the temptation to make small talk. On long distance calls, you're charged by the minute.

4. Save up messages so that more business may be conducted in one call.

5. When possible, prearrange a schedule of times to call so you'll avoid the old cycle of calling, missing the party, having a return call, missing it, recalling, missing, etc.

6. Call different time zones during early morning or evening hours to avoid the high long distance rates between 8:00 a.m. and 5:00 p.m. your time.

7. Use direct dialing for long distance calls.

8. Coordinate and combine calls so that one call serves several needs.

VOICE IS THE KEY

Every time you make or receive a call, you represent your organization and yourself—favorably or unfavorably. The other person or persons can't see you, so the impression you give depends on your voice and telephone manners.

Here are some suggestions for developing a pleasing and effective telephone personality:

1. Maintain a cheerful and considerate attitude toward each telephone call. Boredom and discourtesy are easily recognized and give a poor impression.

2. Strive for these six qualities: alertness, expressiveness, naturalness, pleasantness, distinctness, and helpfulness.

3. No matter how much you know or how sincere you are, you won't impress your clients with a dull, monotonous telephone voice. Therefore, it's wise to cultivate a moderate but lively one that will effectively carry your message over the wire or cable.

4. Remember too, you can't help a client who has to strain to hear or understand you. Here are some ways to improve telephone communication:
   - Keep your lips ½ to 1 inch from the mouthpiece.
   - Pronounce letters, numbers, and names clearly. For an extra margin of safety, spell out names.
   - Organize your thoughts before speaking. This can avoid embarrassing and time-wasting hemming and hawing.

Fig. 6-8. Successful telephone communications require good organization and planning.

- Use complete sentences and concise statements related to the purpose of the call.
- Use words that best express your natural self and message.
- Avoid speech mannerisms and interruptions during the conversation.

PROPER TELEPHONE PROCEDURES

Learning how to use a telephone correctly can improve telephone communications. Small details can make a big difference. For instance:

1. *Identify yourself*, your office, or your department in a few words. Avoid the meaningless, time-wasting, and outdated "hello" and "yes."

   If you're the first person to answer the office phone, identify your office and yourself. If your secretary has referred the call to you, you need only say, "Jones speaking" or "This is Ms. Brown."

   Your own identification given, the person calling will usually respond by giving identification. If this isn't done, try to get the caller's name as soon as possible.

Fig. 6-9. Your voice and telephone manners convey attitude and interest.

2. *Be ready to talk.* Expecting callers to "hang on" while you finish other business can be most annoying.

3. *Return calls.* If you must leave the telephone during a conversation and won't be able to return immediately, say that you will call back and then follow through.

4. *When you are to be away* from the office, *leave word* where you are going and when you expect to return, or leave a number where you can be reached.

5. *Say "goodbye" pleasantly* and replace the receiver gently. The person making the call should end the call.

HOW TO HANDLE THE "REQUEST" CALL

A frequent type of call is the request. Usually such a call requires some reference work. If you must hunt for information, it's wise to handle the call in one of the following ways:

"Just a minute, please. I have it right here."

"Excuse me a minute while I look up that information."

"I find that it will take a few minutes to locate that material. May I call you back?"

Then call back!

WHAT TO DO IF YOU GET A WRONG NUMBER

If you get a wrong number, stay on the line long enough to apologize. Remember, you only waste time by saying, "Who is this?" You'll get results more quickly if you say "Is this Mr. Smith?" or "Is this the McKay Feed Company?" or "Is this 467-7755?"

And, if someone calls you by mistake, be pleasant.

SUGGESTIONS FOR SECRETARIES

Secretaries represent offices and organizations on the telephone just as they do in face-to-face contacts. Thus, it's important for them to put their best voice forward in telephone conversations. How a secretary handles specific situations will vary from office to office. Here are some general guides:

1. Identify the office or department promptly after answering the phone. There is no need to identify yourself, unless you have an executive position.

2. In a tactful and friendly manner, ask who is calling, if necessary. One effective way to get this information is to say: "May I say who's calling?"

3. If the person called isn't in, offer to help. Say, "Ms. Brown isn't in just now. This is Mr. Wilson, Ms. Brown's secretary. May I help you?" Or, "Mr. Jones is at a meeting until 4:00 p.m. May I help you?"

4. If you can't help the caller, offer to take a message. Write down the caller's name and telephone number, the time of the call, the purpose, if given, and the time the call should be returned, if a time has been specified.

5. Be prepared for the often-repeated call that asks for information. Have dates of meetings, names of publications, and other such information on hand. Handle as many routine matters as possible.

◄ Increase Your Listening Power ►

You can listen three times faster than you can talk. The average person speaks at the rate of about 120 words per minute but can listen at 400 words per minute. Comprehension is little affected by the faster speech rate.

It has been estimated that about 98 percent of all learning occurs through the eyes and ears. Seven out of every 10 minutes that you are alive, conscious, and awake you are communicating verbally—9 percent writing, 16 percent reading, 30 percent speaking, and 45 percent listening.

Listening is a critical factor in human development. A person listening deeply undergoes a number of body changes. Electrical activity of parts of the brain connected to the internal ear shows a choppy "activation" pattern characteristic of alerting and attending. Other parts of the brain also have altered electrical activity. When neurons are excited by receptor events (sensory stimuli), their rate of "beating" or "firing" increases. There is both head and eye orientation to the source of sound. A galvanic skin response is present. Blood flow to the head increases, while that to the fingertips decreases. Body temperature may rise. Eyes are bright; posture is expectant. Effective listening is an active, responsive experience.

Sensitive listening is an effective agent for individual change as well as for group development. Listening at a highly conscious and emotional level helps both listener and speaker to become more mature, more open in their experiences, more democratic, less defensive, and less authoritarian. Listening is a growth experience.

Most individuals attend to only about 50 percent of what is said in their presence, no matter how carefully they thought they had listened. Two months later they will remember only 25 percent of what was said. Why? Partly because of poor listening habits.

Studies show that listening comprehension can be increased greatly by using better listening habits. Determine your listening habits by taking the following test. In each of the following pairs, check the listening habit which best describes yourself.

1. ( ) I call the subject uninteresting. I take "target practice" at the topic. I build a shield between the subject and me. How dull can it get?

( ) I'm selfish; I sift, screen, and winnow—always looking for something worthwhile or practical that I can store away in my mind for future use.

2. ( ) I criticize the delivery. "Is this the best they could get? This speaker can't even talk—only reads. Never looks at me. What a voice! Snorts, coughs. Doesn't leave glasses on, etc., etc."

( ) I say, "This person's not great as a speaker; does know something I don't or wouldn't be up there. I'm investing my time in listening, and I'll dig out what is said if it kills me. I'll concentrate on *what* is said and forget about *how* it is said."

3. ( ) I get excited, overstimulated. I have a hard time holding in my comments. I know more than the speaker. I don't like it when my biases and convictions are stepped on. I'd like to "hang" this turkey with this question. I compose neat little rebuttal speeches when I hear a point with which I don't agree.

( ) I withhold evaluation until comprehension is complete. I hear the person out before I make a judgment. I delay my private little strategy session until I've heard all that the speaker has to say on the matter.

4. ( ) I listen for the facts in a speech. Generalizations aren't of as much value.

( ) I try to get the gist of the speech—the main idea. Principles and concepts are the main thing.

5. ( ) I try to make a complete outline of everything the speaker says. I carefully follow the formal rules of outline structure.

( ) I take simple notes, mostly just a listing of facts and principles.

6. ( ) I lean forward with my chin resting in my hand and look straight at the speaker. I breathe deeply and slowly and stay physically relaxed.

( ) I keep my body a little tense. My breathing and heart rate are faster. I don't make an effort to look at the speaker.

7. ( ) Noise, movement, and other distractions in the room bother me; however, I'm sometimes guilty of disturbing others.

( ) I don't create or tolerate distractions. I mentally try to block out everything except what the speaker is saying.

8. ( ) I avoid like poison any communications which are tough, technical, or expository in character. I don't like to be exposed to subject matter in unfamiliar areas. I stay away from the "deep stuff."

( ) I accept difficult and unfamiliar material as a challenge. I enjoy technical presentations; they sharpen my listening skill. I like to explore the new and the unknown.

9. ( ) I am influenced by emotion-laden words; they throw me out of tune with the speaker. Certain words destroy my rapport with a speaker. When this happens, my listening efficiency (learning) may be disrupted for several minutes.

( ) I refuse to let "loaded" words become barriers to listening. I discover which words disrupt my listening efficiency, rationalize them, and eliminate their existence as problems.

10. ( ) I waste the differential between thought speed and speech speed. The average person speaks just about 120 words per minute. I listen at an easy rate of 400 to 500 words per minute. Therefore, I may listen for only 10 seconds and go off on a mental tangent for 50 seconds. My tangent can carry me away completely. I do a lot of mental "island hopping." Much that I hear is old hat anyhow. I don't have to pay much attention to get the message.

( ) I know what activities to engage in mentally in order to stay tuned in on the speech. I try to anticipate the speaker's next point. This can reinforce and nearly double my learning if I'm right. But even if I'm wrong, I'm still learning by comparison and contrast. I attempt to identify the supporting elements used to build the speaker's points. I use a little of the speed differential in identifying this point-support material. I periodically make mental recapitulations of what's already been said. This reinforces my learning to a great degree and turns the differential between thought speed and speech speed into a profit.

ANSWERS: The first item in each pair depicts faulty listening; the second depicts desirable behavior. If you checked the second description in each of the 10 pairs, you're a great listener and certainly a rare person. Anyone scoring less than 5 is badly in need of improvement. Recheck the poor listening habits you now possess. Work on them for six months, then retake this test. Your improved listening should become apparent before the end of this period if you dedicate yourself to this end.

Remember, communications is a two-way process. Both sender and receiver of messages share important roles. The burden can't rest solely on the speaker. The listener must engage totally in the process. This is essential in achieving understanding and reaching common goals. The speaker is merely the initiator of the process; the listener is the sustainer.

Photo / Art Credits

Bill Ballard, North Carolina State University: Figs. 6-4, 6-5, 6-6, 6-8.
Jim McClure, The Pennsylvania State University: Figs. 6-2, 6-3, 6-7, 6-9.
Stan Williams, The Pennsylvania State University: Fig. 6-1.

7 RADIO

◄ Radio as a Communication Tool ►

Before examining radio specifically, let's think about how people learn. What causes individuals to accept a suggestion, a practice, or an idea? What motivates them to act? This change in behavior is called "the adoption process." The process has five stages—awareness, interest, evaluation, trial, and adoption.

Radio and other mass media are most effective at the *awareness* stage, when persons first learn of an idea or a practice, and at the *interest* stage, when they get more information about it.

The most important influences at the *evaluation, trial,* and *adoption* stages are friends and neighbors and personal experiences—particularly experiences with salespersons and other point-of-sales contacts. Nevertheless, radio can play an important role in all five stages. It keeps the practices or ideas before those who otherwise might lose interest. It reinforces the determination to go ahead, to adopt the practice. Research shows that community leaders who first adopt new practices and whose advice is important enough to be followed by others, get much of their information from the mass media.

Although personal contacts and experiences are more influential at the latter decision-making stages, radio through interviews and personal testimonies of success with the product or practice can carry the listener beyond just the *awareness* and *interest* stages.

RADIO IS MANY THINGS

Radio is a personal, even intimate, channel of communication. It brings speakers into the room just as though they were there in person.

Radio reaches almost everywhere—to the home and automobile; the shop, barn, and utility room; the truck or tractor; the feed mill; the machinery dealer; the stockyard—anywhere!

While working at various tasks, individuals can listen to the radio. Homemakers can clean, cook, or shop, and pick up tips from the radio at the same time. Farmers can plow fields, harvest crops, milk cows, hunt game, repair machinery, or feed livestock—all within the sound of the radio. Radio can inform and serve as a companion at the same time.

Fig. 7-1. While working at various tasks, individuals can listen to the radio.

Prepared by **J. Cordell Hatch,** electronic communications specialist and professor of agricultural communications, The Pennsylvania State University.

Radio relies solely on sound, many times on voice alone, *your* voice. And from your voice, listeners decide what type of person you are. You may sound friendly and wise, or impatient and inexperienced.

People get to know you from radio. They know how you sound, what your work is, and how you think, without ever having met you.

Today, radio moves at a rapid pace. Everything is presented in brief segments. Few stations allow much time to slip by without returning to music. Some stations, however, have a "talk format" which allows more flexibility.

Radio thrives on local names, local events, and local situations. Radio gives you regular contact with your audience. Listeners can learn from you what events are taking place now or later, as well as what insects are prevalent and what sprays will control them.

Through tape recordings and direct live broadcasts, radio lets you report from almost anywhere.

This is the fast-moving, locally oriented, personal medium called **radio.**

RADIO CAN'T DO EVERYTHING

Radio can do some jobs better than others. Since it can't show various happenings as television can, it's more effective in reporting management practices, giving market trends, providing nutrition facts, and discussing various public issues, all of which are well suited to radio.

Furthermore, since a listener can't fold up a radio program and file it away for future reference, radio works best when notifying, reminding, or telling uncomplicated stories that are easily remembered. It's most effective when messages are simple and to the point.

HOW RADIO IS ORGANIZED

Stations are licensed by the Federal Communications Commission (FCC) and are expected to broadcast in "the public interest," for "convenience and necessity." This does not mean that a station has to grant free time to agriculture, home economics, or religion, for example.

Never go to a station with the thought that you're doing it a favor by going on the air. Rather, think of the situation as a mutually beneficial arrangement: The *station* provides the facility, and *you* supply the news and useful, practical information. Together you can reach thousands of people. Fortunately, most stations pride themselves on being alert to community needs and are anxious to give good service to the public.

RADIO IS A BUSINESS

Stations stay on the air only if they make a profit. Owners have a right to expect a return on their investment. The sale of air time, in the form of commercials, is their only source of revenue. By comparison with the commercial print media, the ratio of advertising is low.

License renewal with the FCC every three years requires a complete report. A station's performance must match broadcast services promised in the FCC license renewal request. The FCC has the authority to revoke a station's license. Radio has a public trust which it shouldn't violate.

STATION STAFF

Station manager. This person is responsible for the entire station operation, is likely active in many civic affairs, and should be your friend.

Program director. This person develops the programming schedule, influences programming policy, and is your contact relative to regular or special broadcasts.

News director. This person is in charge of news gathering and reporting and is your contact for telephone reports and news coverage of events.

These are the three persons you'll be dealing with most. If the station has a farm director, you'll be working with this person also.

GETTING ON THE AIR

First, get acquainted with the station and its broadcasting format. If it broadcasts primarily news and music, don't expect to nail down a 15-minute program. The station probably prefers something less than five minutes long, maybe just spots.

Be sure you know what kind of program you can produce before approaching the station. When presenting the idea, point out that you can provide news or interesting and useful information to all listeners, that you can give facts on lawns, gardens, insects, and other homeowner problems. Also, indicate that farm residents are a ready-made audience. Above all, the station management will want to know if you can present this information effectively. Can you? Read on.

SETTING THE AIR TIME

A simple survey of your intended audience will

help you determine the best air time. However, even though you have evidence that indicates one time is better than another for your program, the station still may not make this air time available to you. Even though your audience may not be the maximum at the time available to you, it still may be considerable. Through extensive promotion and good programming, you can build a following for your program. Prove yourself, and then you may be moved to a better time.

◄ Preparation and Presentation ►

"Is my radio program fun, or is it drudgery?" We hope it's not drudgery, because listeners can readily tell when your heart isn't in your work.

Competition is just a turn of the dial away, so you must work to keep your listeners' interest. Apply these four principles:

1. Variety
2. Localizing
3. Personalizing
4. Relevance

VARIETY AND BILLBOARDING

Unless your listeners are "hooked" in the first 10 to 20 seconds, you may lose them. An opening to a radio program serves as bait to grab the listeners' attention.

The practice of billboarding is a good way to open your program. Billboarding is the technique of catching and holding your listeners' attention by telling what topics you'll cover. You may use the same technique at the end of the show. For example, if you have a program coming up the next day, you might say, "Tomorrow we'll talk about a new hybrid wheat variety . . . a simple and easy way to fertilize shade trees . . . and how 4-H'er Kathy Jones turned her hobby of gardening into a real business enterprise. Join us again on 'Extension Reports' at noon tomorrow."

Variety helps keep the listeners' attention. Discuss several different subjects. Treat various program segments differently, and keep them short. The program will move along faster as a result. Limit straight talks on one subject to two to three minutes, or less.

LOCALIZING

Your local radio station is primarily interested in local news. Let's look at the news sources that easily lend themselves to localization.

1. Events
2. The week's activities
3. Timely subject matter and problems
4. Experiences of local people
5. Weather and markets

Local events offer a double-barreled opportunity. Promote them before they take place, and then provide follow-up coverage. Taped interviews or reports recorded at the events give your broadcast that needed change of pace (see Fig. 7-2). Localize events by telling how those outside your local area apply.

Fig. 7-2. Interviews taped on location make good radio listening. A portable tape recorder and quality microphone capture on-the-spot interviews for later broadcast.

Your week's activities provide a never-ending source of ideas for radio talks. Who wrote to you? Who stopped by the office? What were their problems? Whom did you visit? What observations did you make?

A tape recorder is an invaluable aid for collecting future program material. Take one with you on county visits. You can even tape interviews with persons visiting your office.

Timely information related to local problems makes good radio listening. And no broadcast beats the successful experiences of local farmers or homemakers in getting practice adoption.

State and national news, including farm news, often affects the counties. You may easily localize wire service news, university releases on research results, and taped recordings of specialists.

Because weather is always changing, it's always sought-after news—whether it's snowing, raining, extremely cold, or very hot. Interpret the weather's effect on crops and farming operations.

Market reports are of critical importance to the producer, and trends are of interest to the consumer. But they are more meaningful when you interpret the effects on local buying and selling.

PERSONALIZING

People like to hear their names as well as see them in print. Interviews rely heavily on names, as well as personalities, and are effective in getting information across. Interviews also help to personalize your program. Work in names as often as possible without becoming a conspicuous name-dropper.

RELEVANCE

There can be variety, localization, and personalization, but unless the content of your radio program is relevant to the interest and needs of the station's listeners and your target audience, the whole process becomes an empty exercise.

The content should address basic human needs as well as specifics your listeners may want to "have," "save," "be," or "do," such as money, time, happiness, comfort, safety, success, recognition, respect, etc.

Try to determine what will motivate your audience. What do these people want? What will benefit them? What changes in knowledge, attitudes, or behaviors do they seek or can you stimulate? Listeners must see some personal payoff in what you say. Sell benefits! Point out very plainly the relevance and payoffs rather than have your listeners search for them. This is a good way to start your broadcast.

SUGGESTED FORMAT: FIVE-MINUTE SHOW

1. A good opening and billboard: 10 to 20 seconds.
2. "News from around the county": one minute. Describe the current work of farmers and homemakers.
3. Feature material: two minutes. This could be advice, information, or an interview with a farmer, homemaker, field worker, or specialist.
4. "Farm or home calendar": one minute. Announce coming events and tell where they will be, when they will be held, and who will be there. In each instance, describe the program in an inviting way that will make listeners want to attend. Couple the announcements with nuggets of subject matter information.
5. Reminders, news summaries, a preview of your next broadcast, and a friendly close: 40 to 50 seconds.

Between the program's segments, give your listeners a mental bridge. Don't stop one topic abruptly and immediately swing into the next. Use short, natural transitions to bridge the gap.

Five-minute programs are flexible and versatile. You can change the format easily. You might give a straight talk one day and interview a farmer or a specialist the next. News features or combinations of items provide other possibilities.

With a daily five-minute spot, you can build a tremendous audience if you offer concise, timely, interesting, relevant information.

GETTING ON AND OFF IN TIME

Occasionally you may run overtime on your program. It's worse, however, to run out of material. You won't have this trouble if you keep short fillers handy. Time your presentation before going on the air so that you'll know when to start wrapping up. Discussing the content of your next program—billboarding—is a good way to ease out. Don't say: "I see by the old clock on the wall . . . , etc."

Theme music helps to identify your program at the beginning. And it often acts as a time cushion or "pad" at the end of the program.

DON'T FORGET PUBLIC RELATIONS

Your program helps the station do a better job of serving its diversified audiences. At the same time, the station personnel can help you do a better job. Ask for their ideas and constructive criticisms. As a

Fig. 7-3. Keep your eye on the clock, but don't talk about it on the air.

goodwill gesture, invite them to agricultural and home economics functions as your guests. This is good public relations. Always remember you are your own best "PR" person. So relax and let the "real you" come through on your radio presentations.

A FEW DON'TS

● Don't beg, plead, or threaten in an attempt to get or keep free air time. A good program will stand on its own merits.

● Don't use free air time to promote a project or an idea for which you'll buy newspaper space. If the project has an advertising budget, allot some to radio. How do you think the station manager feels when he/she sees a paid newspaper advertisement the morning after you used free time on the station to promote this same project?

● Don't promote anything related to games of chance.

● Don't "plug" commercial products.

● Don't tinker around with station equipment.

● Don't talk with announcers on duty unless asked to do so. They and other people at the station have busy schedules. Be friendly, but don't hang around needlessly. Be careful what you say in any studio; you never know when a microphone may be "live."

◀ Writing for Radio ▶

Know why you're using radio. Before you write one line of copy or say a word on the air, decide exactly what you want your listeners to **know, feel,** and **do.** How do you want them to behave, react, change? List the ways as specific message objectives.

Decide on one basic, timely idea for each story. It may be part of a larger idea. Collect and organize logically all related information, facts, data, and substantiating evidence which will make your **message** believable and acceptable to those who hear it.

Get your listeners' attention. Use an unusual fact, an interesting idea, a thought-provoking question, or a challenging statement. Arouse interest, curiosity. Your lead must catch the attention of even the most casual listener. The first two sentences of your story or spot are the most important. If you don't get the listeners' interest in the first 10 seconds or so, chances are you won't at all. You can use a good line more than once or in more than one place. Try reading this: "Feeding immediately following the flowering period of many shrubs is becoming a common

and recommended practice." Confusing? You bet! But use the same content for this kind of attention-getting lead: "Been wondering about fertilizing your shrubs? Well, wonder no more . . . shrubs do need fertilizer, especially right after flowering. For many shrubs, that's *now.*"

Use mass audience appeal—don't exclude listeners. Even though your **message** may be primarily intended for a relatively small audience, present your information in such a way that it will be of interest to many listeners. Which of these two versions has the broader audience?

Poultry Raisers—egg production is highest and profits are greatest when layers receive 14 hours of light per day.

Or:

Hens don't lay eggs in the dark! They need 14 hours of light each day for top production. This pays off in two ways: higher profits for the poultry raisers . . . and lower egg prices for us all.

Instead of talking "to" farmers, talk "about" them. Tell what they do and why, and how this is important to non-farmers.

Give the source of your information logically, naturally. Listeners are more likely to believe what you say and write if it is credited to an authoritative source. If the source is you, then try to indicate on what basis. Other sources should be identified, but not belabored. Handle attribution as if you were "telling" the information to a friend. For example:

Not this:

Mulching provides the best method to insure an adequate supply of moisture around shrubs, Penn State extension horticulturist Craig Oliver says.

But this:

Shrubs need a steady supply of moisture. And according to Craig Oliver, extension horticulturist at Penn State, mulching is your best bet.

In radio, attribution generally precedes the statement, whereas newspaper style usually has the source following the statement.

Make your story easy for your listeners to follow. Remember, radio is the "hearing" medium. Each word, each idea must logically fall into place so your listeners can get your message easily and accurately. Your story must have a "liquid" or flowing quality. Words like "however," "but," "on the other hand," "or," "and," "furthermore," "therefore," "so," "well," "then," and "likewise" act as "mental steering wheels" to guide your listeners' thinking. Use these words to start sentences.

Fig. 7-4. Gesturing and smiling, even to an empty studio, adds personality and vitality to your broadcasts.

Use simple and understandable words and terms. Use as few technical words or terms as possible. Find the simplest word that will carry your meaning. In each of the following instances, the word after the dash is better than the one before it: "utilize—use"; "purchase—buy"; "procure—get"; "accomplish—do"; "eradicate—wipe out"; "contribute—give." If you use a new or unfamiliar word, explain briefly what it means.

Break into syllables names and words hard to pronounce or often mispronounced.—For example: "dacron" (DAY-KRON), "debris" (DAY-BRE). In some cases, spell the word for clarity and emphasis: "Sinox—S-I-N-O-X."

Use short, easily read, easily understood sentences. Keep a period at your fingertips. If you use a long sentence, follow it with a short one. This will permit you to catch your breath. Strive for a comfortable rhythm. Remember, the listeners must comprehend instantly. So don't get yourself or your listeners bogged down in a mass of complex details or instructions, lengthy formulas, or involved procedures. They are not suited to radio.

Use simple, direct "picture language." Avoid words which dilute meaning. Use examples and comparisons with which the listener is familiar. Paint word pictures, such as "icing so smooth you hate to put a knife into it," "little orphan pigs, little four-footed feed factories," "mold that looks like a light coat of flour on the leaves." Use action verbs and a minimum of adjectives. Don't judge the information for your listeners. Present the facts. Descriptive detail allows your listeners to decide for themselves whether the information is useful or whether the event is "big" or "exciting." How big is "big"? What is "exciting"? Good word pictures speak for themselves.

Use contractions. The copy will read more easily and sound better. For example, say, "Here's advice from . . ." instead of "Here is advice from. . . ." Use "don't" instead of "do not." But beware of contractions such as "could've" and "would've." They sound like "could of" and "would of," which are improper grammar. At times, for emphasis, you won't want to use the contracted form, such as "For safety's sake, *do not* leave chemicals within the reach of children."

Write words and symbols so they are easy to read and understand. These words and symbols are written after the dashes as they should appear in radio copy: lbs.—pounds; 50°C.—50 degrees centigrade; 1 m.—one meter; %—percent; 98—almost one hundred; N—nitrogen; $1,425,625—about 1½ million dollars; 56,875—56 thousand 875, or nearly 57 thousand.

Repeat or re-emphasize the parts of your message that your listeners may have missed when you first gave them. This can be done as a summary or as a call for action and may include addresses, amounts, dates, titles of publications, sources of additional information, or key points that need stressing again. An offer might be repeated as follows: "Again, to get your copy of the *Home Gardener's Handbook,* send your name and address to: Handbook . . . WPSX . . . University Park, Pennsylvania 16802."

Test your writing by reading it aloud. This is the best way to catch poor phrasing and tongue twisters. If you can read your copy easily and smoothly, chances are others can too. Check for readability and listenability while you are writing, as well as after completing the story.

MECHANICAL HELPS

Have your script typed clearly in the usual manner (not in all capitals), and double or triple space so it's easy to read. Avoid splitting sentences at the end of pages and words at the end of the lines. Write words out—don't abbreviate.

Use 8½- × 11-inch mimeograph or other soft fiber paper. Never use onion skin or bond paper; it rattles. Use a separate sheet for each story. Time your copy by reading aloud, or figure at the rate of 130 words per minute.

Number all pages. Underline words you wish to emphasize. Mark places where you wish to pause. Note pronunciation trouble spots. Leave pages loose—don't clip or staple them together.

USING SPOTS

If you listen carefully to your favorite radio station, you'll find that most of its commercials and public service announcements are 20, 30, or 60 seconds in length. The commercials are packed with sales power, while public service spots are designed to inform and change behavior.

Spots can be presented live or taped. And, because they're short, they're more apt to be repeated during a broadcasting day than are longer programs. The writing techniques described earlier apply equally well to spots and stories. But, the short duration of spots calls for especially tight writing and concept simplification.

◀ For Better Delivery ▶

Your attitude or psychological state greatly affects your radio delivery. In fact, the right attitude can compensate for other shortcomings in delivery. Keep these four points in mind:

1. **Think the thought.** Regardless of the topic or idea, think about it, see it, feel it. Visualize the insect pest you are describing. Taste that suggested low-calorie dessert. Be impressed by that new milking setup.

2. **Think the thought through to the end.** Read or speak by phrases or logical thought units. Know how the sentence will come out before you start it. Keep half an eye on the end of the sentence while you are reading the first part. This will add smoothness to delivery and will aid you in interpreting the meaning of the phrases as parts of the whole idea.

3. **Use your body.** A relaxed body helps to produce a relaxed-sounding voice. Do a few exercises just prior to going on the air. Physical activity reduces tension, but don't be out of breath. A little tension, however, is useful in maintaining mental sharpness.

The voice reflects your state of mind and body. During humorous lines, smile—your audience will "hear" the smile in your voice. To emphasize a point, use your hands—much as you would if you were talking to a farmer sitting across the desk from you. Frown, shake your head in disbelief, count off on your fingers—all these come through and help you "feel" the material and help your audience get the message.

4. **Talk to someone.** Good on-air people keep reminding themselves that the purpose in speaking is to convey an idea to someone else. Talk to your listeners, *not the microphone*. Talk to

them with the realization that they've never before heard what you're saying and they may never hear it again. You must get your meaning across to them right now.

These four rules of delivery are important, but you shouldn't be satisfied until you have achieved proficiency in speech and delivery techniques.

A national survey indicates that the biggest faults in educational programs are lack of enthusiasm and poor voice quality and characteristics. It's not enough, however, to know that you should sound natural, friendly, warm, at ease, interested, and enthusiastic; you must enunciate clearly and correctly; pace yourself well (not too fast or too slow); and avoid speech mannerisms.

It takes practice and conscious effort to perfect the art of radio speaking. Record and carefully critique all your practice sessions and broadcasts. This will make you more aware of your strengths and weaknesses. Also, it will create a heightened sensitivity to all those factors which can improve your broadcasts.

(The material covered in Chapter 6 should be helpful too.)

SUMMARY—DELIVERY TIPS

Speak clearly in your normal, conversational, friendly tone. Think of yourself as talking to one person close by, not to a big crowd. Your aim should be to *talk* to your listeners, not to read to them. Project your personality. Sell your listeners on the points you're making. Be persuasive. Enthusiasm and sincerity will help to convince them you believe what you're saying.

Talk at natural speed, but change occasionally to avoid monotony. Vary the pitch and volume of your voice to get variety, emphasis, and attention. These changes will sound artificial, however, unless kept consistent with subject matter content.

Control your breathing to take breaths between *units of thought;* otherwise, you'll sound choppy. Avoid dropping your voice when it sounds unnatural to do so. Make your voice pleasant; a smile on your face will put a smile in your voice.

Watch enunciation and pronunciation. Do you enunciate each part of a word clearly? Or do you slur certain syllables? Do you say "temp-a-ture" for "temp-**er**-a-ture"? "Prob-ly" for "prob-**ab**-ly"? "Git" for "get"? "Jest" for "just"? Don't say "sewin'" or "cookin'"—sound the final "-ng's." And "often" is pronounced "**of'n**"; the "t" is not sounded.

Practice your radio presentation with someone listening who will be frank in criticizing your delivery as well as the content. Also, listen critically to tape recordings of your delivery as often as possible.

Be Chatty.
Be Yourself.
Be at Ease.
Be Enthusiastic.

◀ Radio's Working Tools ▶

There's a wide assortment of tools and techniques designed to make your radio programs more effective and easier to do. It's important that you look upon these as aids, "helps," or "little assisters," not as "monsters" or obstacles with which you have to contend. Let's examine some of the items you can put to work for you.

TAPE RECORDERS

The tape recorder is undoubtedly the most important broadcasting development in years. Invented in Germany during World War II and called the "magnetophone," the first tape recorder was manufactured in the United States in 1946. Since then, models and designs have become so numerous that to select the one "most suitable" model is almost impossible. We will try to touch on some of the more important considerations.

Speed. This has to do with how fast or slowly the tape is pulled past the record head. Common speeds, measured in inches per second (i.p.s.), are 1⅞, 3¾, 7½, and 15. High-speed duplicators run at 30 and 60 i.p.s. *Radio has accepted 7½ i.p.s. as standard* (see Fig. 7-5). Most recorders now on the market are multispeed; that is, they may have both 3¾ and 7½, or 7½ and 15, or even 1⅞, 3¾, and 7½.

For purposes other than radio, these faster and slower speeds may be quite useful. However, in selecting a recorder for radio use, be more concerned about the accuracy of speed than the number

Fig. 7-5. Common tape recording speeds range from 1⅞ to 15 inches per second, but 7½ i.p.s. is the standard in broadcast radio.

of speed settings. Furthermore, the speed should be constant. Any variation from faster to slower and back again, however slight, is irksome. A tape recorded on an imperfect machine may be rejected by the station.

Timing tapes allows you to check speed accuracy, and a listening test with piano music played back usually accents speed variation. There's no easy cure for speed problems. Simply don't buy or use for radio purposes any recorder with such ills.

Track width. All the early model recorders were equipped with full-track heads. Today, however, you are more likely to see machines with half-track, quarter-track, or combination heads, as in several of the stereo units. The more costly professional recorders have optional full-track heads, and one of these is what you should buy if you can afford it. Be sure to specify full-track heads. Half-track recorders may be used for radio if you always start with clean or demagnetized tapes.

Width refers to the vertical gap width in the record and playback heads. The recording tape is ¼ inch wide. A full-track head records across the entire width, or ¼ inch; a half-track head records across one-half the width, or ⅛ inch; and a quarter-track head uses one-fourth the tape width, or ¹⁄₁₆ inch.

You can imagine what would happen if you recorded a radio program on one-half the tape, and something entirely different was left on the other half. At the radio station the tape is played back on a full-track machine, which picks up and broadcasts both sources, both tracks, simultaneously. Trouble? Yes, indeed!

Power requirement. If you're going to have only one recorder, it should in all likelihood be an AC-powered model. Why? It's more dependable, its quality is much higher, and its audio output is greater.

Some models may be used as small-group public address systems. For this purpose, have a radio serv-

icer or engineer install a motor cutoff switch so the motor isn't running, generating heat and noise, while the electronic circuit is serving as an amplifier for your voice.

The battery-powered recorders have become more popular and dependable in recent years. The more expensive reel-to-reel and cassette models have good quality and are easy to use. Their value for on-location recording makes them a practical investment. There's no problem in dubbing from these units to the AC models. Together they provide great flexibility in programming. If at all possible, buy a battery-powered recorder as a complement to your AC-powered machine. Acceptable cassette units are available at $50 to $300. Better reel-to-reel units may cost $200 to $300; the best ones, $2,000 or more.

Fig. 7-6. Professional quality is available in multi-channel decks and in battery-powered cassette and reel recorders.

Inverters. These allow you to operate AC-powered equipment from an auto battery. They convert DC energy to AC, making it possible to operate an AC recorder or other equipment with battery power.

MICROPHONES

Put simply, microphones are devices that change acoustical energy into electrical pulses. Although the microphone principle of sound reproduction hasn't changed in years, new developments and refinements have been incorporated in microphone manufacture. The five basic types of microphones are:

1. *Condenser*—excellent frequency response, low distortion, expensive.
2. *Ribbon*—very sensitive to movement, good frequency response, expensive.
3. *Dynamic*—sturdy, unaffected by atmosphere, good frequency response, medium priced, your "best bet."
4. *Crystal*—widely used, low cost, good response, affected by temperature and moisture, fragile, provided with most recorders.

5. *Carbon*—budget mike, poor response, not widely accepted.

Each mike has its own set of characteristics with respect to pickup pattern, output impedance (high or low), frequency response, and output level. These characteristics determine the type of microphone best suited to your recorder and need.

RECORDING TAPE

Tape consists of a very thin layer of iron oxide emulsion cemented to an acetate or mylar base. The recorder generates magnetic pulses, somewhat as a horseshoe magnet does. These pulses arrange the iron particles on the recording tape into "sound patterns."

The thickness of base and emulsion determines the amount of tape (number of feet) which can be wound on a reel. The tape length determines the playing time, as time is measured in terms of inches per second. For example, at 7½ inches per second, it takes 30 minutes for 1,200 feet of tape to pass the record head. The standard thickness for radio is 1½ mil. Thinner, long-playing tapes, ½ or 1 mil, may not work well on some professional machines. Compared to the 1½-mil tape, they're more fragile, harder to edit and use, and subject to stretch and print-through. So stick with the standard thickness (1½ mil), unless you need tape for longer recording sessions.

TAPE EDITING

The magnetic sound patterns on recording tape are similar to the printed word. Just as words, sentences, and paragraphs on a printed page can be edited, deleted, or rearranged, so can words, sentences, and paragraphs on tape. With a little practice you can become quite adept at structuring the tape content to fit your particular needs.

The irrelevant, the unwanted, the distracting have no place in any radio presentation. Remove them by editing. Often it's necessary to shorten a program. Do this by editing out the weaker portions, keeping the "meaty" content. Experienced tape editors have perfected the art to the point that they can make singular words out of plurals by cutting off the "s" sound.

There are two editing methods: (1) cutting and splicing the original tape and (2) dubbing from one recorder to another—electronic editing.

Cutting and splicing. This is the most frequently used method, principally because it's more precise than dubbing. The first step is to remove the head cover and locate the playback head. Identify and mark the magnetic gap on top of the head. Next, listen to the tape carefully, perhaps two or three times, before you do any cutting.

Decide what you want to take out and how you want the finished tape to sound. Use a grease pencil on the backing side of the tape to mark the starting and ending points of the portion to be cut. Place each mark in the short silent space between the words you want to keep and the segment you want to take out.

With a razor blade or a pair of scissors, cut to the inside of the marks on the outtake portion. Lay the outtake aside; don't discard it until the finished product is to your satisfaction.

The tape is now cut, and you have two loose ends with marks on them. Superimpose one mark over the other, and with scissors make a diagonal cut through the marks. A little of the grease mark will be on each end of the tape. Remove it with a piece of cloth or a paper tissue.

Now, on a smooth surface, butt the two angle cuts together with the backing side up. Hold them in place with two fingers while you put a ½- to ¾-inch piece of special editing tape over the joint. (Never use ordinary cellophane tape except to repair emergency breaks.) With your fingernail, press out air bubbles. Then lift the tape from the work surface.

With scissors trim the splicing tape which extends over the edges, taking a tiny bit of the recording tape. The edges of the splice must be smooth or they'll catch and be rough going through the tape guides and head assembly.

Your editing job is made much easier and faster if

Fig. 7-7. Many of these older, open-reel recorders are still in use, and cutting and splicing is relatively simple with this type of machine. Use a grease pencil to mark the starting and ending points of the part of the tape to be eliminated.

you use a grooved splicing block. The block allows you to make precise cuts with a razor blade (see Fig. 7-8). The splicing tape is a little narrower than the recording tape and is placed over the cut so that it doesn't extend beyond the edges of the recording tape. If it's used carefully, no trimming is necessary.

Dubbing. If fairly long segments are to be taken from a tape, this may be done easily by dubbing from one recorder to another. Connect a patch cable from the output of one to the input of the other. Make a test run to check quality and set levels.

If your recorders have pause or edit switches, your job is made considerably easier. Cue both tapes, and then start them simultaneously. It's that simple. It usually helps to locate and identify the closing phrase so the tape can be stopped before reaching unwanted content. After the last sound of the phrase passes the playback head, hit the edit or pause switch or quickly turn down the volume. Practice a little, and you'll have no trouble.

The dubbing method is particularly useful for in-

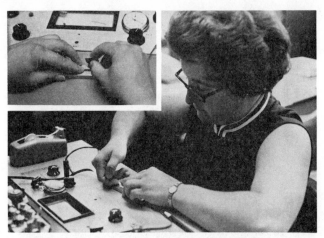

Fig. 7-8. Splicing block and razor blade (inset) make tape editing easy. Narrow splicing tape joins the cut ends together for a strong, quick edit.

serting interviews and other previously recorded material into a program that you're taping. But precise editing is best achieved by the cut-and-splice method.

◄ Better Recording Techniques ►

More and more agriculturists and home economists are now recording their programs prior to air time, rather than doing them live. This isn't surprising. Tape recordings offer several advantages and conveniences—flexibility being the principal one.

When you use the tape recorder as your broadcasting mode, instead of merely being a presenter, you take on the added responsibilities of an engineer—responsibilities for which you may have little or no training and perhaps even less confidence. The equipment is really not as complicated as it may appear. With a little practice, you can make good recordings.

If the station complains about a tape and says the quality isn't up to its standards, your first impulse is to blame the recorder—yours or theirs. The equipment should, of course, be maintained in top condition at all times; however, experience has shown that just as often the problem isn't equipment but human error.

Proper recording and microphone techniques become especially important with the low-cost, non-professional equipment used widely today. Noisy recording areas and poor acoustics, commonplace in

most offices, increase your chances of making poor quality recordings.

Difficulty is likely if your recorder isn't compatible with the station's. Most hometype recorders are equipped with stereo or half-track recording and playback heads. The professional machines used at the station usually have full-track heads. The two may sound incompatible, but they're not—that is, if you take a few precautions. The following procedures help and are basic to good recording regardless of the types of heads and models used:

● *Start with a clean tape.* Have the station erase the tape on its full-track machine or on a tape demagnetizer (see Fig. 7-9). As a poor alternative, you may run the tape through your half-track machine on record with the volume turned all the way down. Both tracks on the tape *must* be cleaned. After this, record in one direction only.

● *Check the recorder head alignment and speed* (7½ inches per second). Most radio engineers can make these checks for you with an alignment and timing tape.

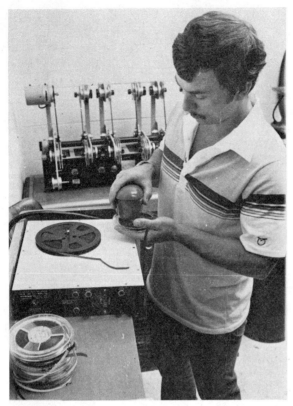

Fig. 7-9. Tape erasers (demagnetizers) remove previously recorded signals and should be used prior to recording.

● *Periodically clean* the recorder heads, the capstan, and the rubber pressure roller with a cotton swab dipped in alcohol.

● *Turn the tone control to full treble position* if it is employed when you are recording.

● *Turn the monitor speaker off* when you are recording with a microphone.

● *Record at the maximum volume at which no distortion occurs.* For a half-track tape to have sufficient volume when played back on a full-track recorder, a high recording level is necessary. The higher volume on the recorded track helps cover up any noise that might be present on the unused half of the tape going over the full-track head.

   If your recorder has a VU meter, adjust the volume upward to the point where the needle just touches the red on the loudest passages. The sound should be clear, strong, and undistorted when played back on your machine. Do a few test runs.

● *Follow the recorder manufacturer's operation and maintenance procedures.* Read and follow the manual's instructions.

Remember, your recorder is an instrument; it takes a lot of practice to operate it well. The more you use your recorder, the better you'll be able to use it.

Treat your recorder as a valuable instrument. Don't bang it around; underneath that rugged shell are delicate operating parts. Don't haul your recorder in the trunk of your car. If possible, place it on the seat, on a cushion, or on a piece of foam rubber.

Your recorder will function better if it's warmed up first. Run the tape for a couple of minutes before starting to record. This is especially critical in the winter when the recorder has been in a cold car for several hours. If possible, bring it into a warm building several hours before recording time.

MICROPHONE USE

1. *A good microphone is necessary* for good recordings. Fuzzy, muddy, distorted sound, or no sound, often is traceable to a faulty microphone. Try your microphone on another recorder or try another microphone on your recorder. This narrows down the problem.

2. Under most recording conditions, room acoustics are far from perfect. This makes *proper microphone placement* imperative. Try several different microphone placements. Select a room which is free from noise and echoes. A "dead" room is preferred to one producing the "rainbarrel" or "bathroom" sound which results when audio waves bounce off hard, flat surfaces. Books, draperies, acoustical tile, angles, and carpeting absorb reflected sound.

3. Hold the microphone 6 to 10 inches from your mouth and at a 45° angle to the direct line of speech (see Fig. 7-10). This will help

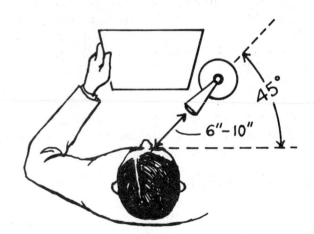

Fig. 7-10. For best results, place the microphone 6 to 10 inches from your mouth and at a 45° angle to the direct line of speech.

prevent "blasting" and distortion of "p's" and "s's" and will diminish the effects of poor room acoustics. The optimum distance and position will vary with different microphones. Experiment! In any case, sit or stand in a comfortable position.

4. *Place the microphone as far from the recorder as possible.* A sensitive microphone may pick up the recorder's motor and tape transport noise. To avoid recording vibrations and motor noise, never put the recorder on the same table as the microphone.

5. For minimum noise and hum, *do not use a recording area lighted with fluorescent tubes.*

6. When you start recording, be prepared to keep at it. Stopping and starting usually results in distracting noises and varied recording level and voice quality.

7. Don't speak for 10 seconds, immediately before and following recording. This gives the tape a clean, silent leader and close.

8. *Speak in your normal conversational tone and volume.* Don't whisper and don't shout.

9. When talking, always maintain the same distance from the microphone.

10. When clearing your throat or coughing, be sure to turn away from the microphone.

11. Don't blow into the microphone to see if it's "on." Tap it with your finger instead, or simply speak into it.

12. Remove all paper clips and staples from your copy before starting the broadcast. Keep your notes still.

13. Avoid tapping your pencil or clicking your ballpoint pen. Caution your guest about this.

14. When holding a microphone, remove your rings so they won't hit against it.

15. *Check tapes periodically* for tears, bad splices, chipping emulsion, damaged edges, brittleness, and other defects.

TIPS ON STORING RECORDING TAPE

1. Store reels of tape in containers. The original box protects a tape from dust and physical damage to its edges. For periods of long storage, metal cans sealed with adhesive tape are best. If the tape came in a sealed plastic envelope, leave it there until you're ready to use it.

2. Wind reels of tape loosely. Store "on edge," not flat. Avoid stacking, as the weight may warp the plastic reels or damage the edges of the tape.

3. Store away from radiators, heating vents, hot water pipes, and windows. Don't leave tapes on a car seat or on the ledge behind the back seat, exposed to the sun's damaging rays.

4. Store away from even slightly magnetic fields, such as those created by electric motors, magnetic tools, TV sets, or computers.

5. Store at an ideal temperature (72°F.) with an ideal humidity (50 percent).

6. If tape is exposed to extreme temperatures, such as in mailing, allow several hours for it to return to room temperature before using.

7. Record tape masters at a low level. They will maintain their high sound quality for a longer period than heavily recorded tapes.

8. Use mylar-base recording tape. It lasts longer and has a greater shelf life than does acetate-base tape. Mylar doesn't dry out as fast, and it maintains a stronger bond with the ferrous oxide emulsion.

9. Use thick tape, which stores better than thin or long-play tape. The problem of magnetic print-through is not as severe on thick (1½ mil) tape as on thinner tape.

◀ Better Radio Interviews ▶

Good radio interviews don't "just happen." It is your responsibility to make your interviews good, not the responsibility of the person you are interviewing.

YOUR GUEST

Ask yourself: Is the person's message important to your listeners? Is it informative, thought-provoking?

Does it create a desire for further study or exploration?

Is your guest the best available choice? Can and will the person speak with authority? Will the material lend itself to the interview technique, or will it be more effective as a "straight talk"?

Once you decide, get some information about the person's background and training, and become familiar with his/her field of work. Then arrange a time for the interview.

PLANNING THE INTERVIEW

Meet your guest and do your best to generate enthusiasm about participating on your program. If it's a first appearance, take time to explain your program, emphasizing why his/her message is important to your listeners. Discuss radio procedure and scheduling, and assure your guest that you'll be there to "save the day" should anything go wrong.

Discuss your guest's work in general, but focus on a specific area, emphasizing two or three main points. List the points in their natural sequence, and suggest a time block for each.

Now you're ready to develop the interview. Make sparse notes only, and encourage your guest to do the same. Explain the importance of being natural and why you don't want set questions or a word-for-word script.

As you visit with your guest, list "lead" questions that you can ask if necessary to move the interview along smoothly and on schedule. Also make brief notes of the information your guest will give in answering the lead questions. These will help you if he/she should "freeze" or forget some of the important points.

Don't ask too many questions. Encourage your guest to talk, and use questions or comments only to draw the person out to clarify a point.

Tell your guest what you're going to ask. Try out your questions before the program goes on the air. But don't do the entire interview. A completely rehearsed interview usually loses some of its sparkle and spontaneity. A little warmup is all that's needed.

Write out a short opening and closing. In that way, you'll introduce your guest quickly and easily bring the interview to an end on time.

ON THE AIR

Prepare your listeners for your guest and the message to be presented. In a field interview, get your listeners "mentally on location" with you and your

Fig. 7-11. In interviewing, have a friendly, "easy" attitude toward your guest.

guest. Remember that they meet the speaker through voice only. Keep your introduction short—don't give a life history. Briefly set the stage, by giving information that has some relation to the guest's message and by explaining what you think will be of special interest to the listeners. At all times try to keep your guest relaxed.

Don't introduce your guest with—"Would you tell us who you are, where you live, and what you're doing here?" This is information you should give. Convey the feeling that you have gotten to know your guest and have discovered that he/she has some really interesting information to share.

Open the interview with a direct question and one that will require your guest to take the lead in the discussion. Ask questions that can't be answered by "yes," "no," "I don't think so." Avoid obvious questions; they waste time, and make you appear incompetent as an interviewer.

Begin your questions with "who," "what," "when," "where," "why," or "how," relying heavily on "how," "what," and "why." The purpose is to frame a question in such a way that your guest can't answer it with a "yes" or a "no."

If you begin a question with "Do . . . ?," "Did . . . ?," "Are you . . . ?," "Is it . . . ?," "Were you . . . ?," or "Have you . . . ?," you automatically invite a "yes" or "no" reply. Too many "yes" or "no" replies from your guest force you to do most of the talking.

Another, and even more troublesome, way of getting into the "yes" or "no" reply is to give the information yourself, and then ask, "Isn't that true?" A few of these statement-questions, and you'll be interviewing yourself.

Don't ask compound questions—that is, two questions in one. Keep your questions simple and straightforward. Two questions in quick succession are better than combining them into one.

Let your guest talk. Ask a question only when needed to bridge, to bring out new information, to keep the guest on the subject, to move the discussion along, or to "rescue" the interviewee.

Try to keep your guest mentally on the farm or in the home or place of business. Don't ask surprise questions. Find out what your guest has done, where, why, how it was done, and what the results were. Stress the pronouns "you" and "your" in your questions. Ask your guest to talk in terms of "I," "my," and "mine." Make a special effort to show that you're genuinely interested in what your guest has to say. Maintain eye-to-eye contact and a warm, close feeling; nod, smile, give positive, supportive feedback. This will demonstrate your interest and boost your guest's confidence.

Keep your eye on the clock, and move the interview along according to the time blocks so that you can include the information as planned. Call your guest by name from time to time during the conversation. This will be helpful to the listeners who tuned in after your introduction, and it will convey your interest to the interviewee.

End the interview on time and in a gracious, friendly manner. Don't let your last question be "Do you have anything to add?" or "Have we covered everything?" Leave the listeners with a feeling that they have met your guest and that he/she has made an important contribution. Make your guest feel like coming back for another visit.

EXTRA DIVIDENDS

The interview technique is neither simple nor easy to master. Take time to get acquainted with your guest and his/her field of work. The better you know the person, the easier you can ask purposeful questions. You will be more natural, your guest more comfortable, and your listeners more interested.

Be interested in your guest as a person. Be knowledgeable and enthusiastic about the message. Your listeners will know if you aren't. Allow a "cushion," or a little extra time, in your program so that you can ask off-the-cuff questions if they come to mind. Frequently they are more interesting than the planned ones. Such questions will come if you follow what your guest is saying.

Avoid unnecessary comments and expressions such as "I see," "un-huh," and "yes." They are dull and usually unnecessary.

Remember, a good interview requires advanced planning and preparation, just as any good on-the-air performance does.

◄ Reporting by Telephone ►

The telephone is one of the easiest and fastest ways to get news and information on the air. Your report can be carried live or recorded for later broadcast.

WHY USE THE TELEPHONE

The immediacy and atmosphere generated by telephone reports help listeners feel that they are at the event. The communicator who uses this method of reporting also has a better chance of hitting prime-time newscasts. As with spots and public service announcements, the report must be crisp and tightly worded. The brief style that is essential may contribute to a better overall reporting job.

The voice quality of telephone reports is *not* always the best, but the immediacy of the report more than offsets this. The "on-the-scene" approach to news reporting has high listener appeal.

WHEN TO USE TELEPHONE REPORTS

Report news *when* and *where* it's made. If there's a telephone available, use it. Your local station wants news and will be happy to get your call.

You can provide up-to-the-minute coverage of 4-H Club events, field trips, livestock shows, sales, contests, fairs, and other activities of interest to the public.

Sudden and dangerous changes in weather conditions, such as unexpected frosts, deep freezes, hail, snow, or heavy rains, and their implications, merit telephone news reports. Insect or disease outbreaks may deserve fast and special treatment. The airways are as close as your telephone.

Radio telephones are now being used at remote locations and on farm tours. On a tour, for example, comments and discussion of farms and fields are transmitted via radio telephone to the local radio station which broadcasts them. Persons on the tour hear everything on their car radios as they ride along. Even persons not on the tour enjoy the tour guide's descriptions—something of a gimmick, but it has worked successfully.

MAKING ARRANGEMENTS WITH THE STATION

Discuss the possibility of telephone reports with local radio station representatives. Are they interested in receiving such reports? Do they have the necessary equipment? Do they want advance notice? What is the best time of day for them to receive your call? Whom should you call at the station when you have a spot news item? Is the station willing to pay for a collect call? (There may be times when your office should pay any long distance charges.) How is the station likely to use your report—as a news feature, item on the farm show, part of a disc jockey show? Does the station have a preference as to length?

A feature may run three or four minutes. An announcement, news item, or statement may be 60 seconds or less.

ORGANIZING AND PREPARING THE REPORT

Prepare and organize your notes before calling the station to make your report (see Fig. 7-12). If it's a report from a fair, a list of county contestants and placings will be most helpful.

You might use this or a similar format:

Today at the Pennsylvania Farm Show here in Harrisburg, three Centre County 4-H Club members won top awards.

Eleven-year-old Becky Glover of RD 1, Coalsburg, led her 1,100-pound steer into the winners' circle after beating out some 40 competitors in the Baby Beef Show. "Midnight," as she calls her jet-black Angus, will be sold to the highest bidder on Friday. Last year's record price was $5.12 a pound.

Becky attends State College Junior High and is the daughter of Mr. and Mrs. R. B. Glover.

In the Shropshire lamb division, Susan Sparks of Millheim placed third and won a blue ribbon in fitting and showmanship.

Carl Snyder of State College won the barrel race on his palomino pony in a time just two seconds short of the 4-H Horse Show record. . . .

Notice first that we told where the report was coming from, then gave a summary opening statement before getting to specific details. Usually it's better to let the news announcer introduce you, rather than opening by giving your own name and title.

Completeness of details, results, highlights, human interest, impressions, color, and humor are governed by the time allowed you on the air.

Sign off in this fashion: "This is County Agent John Smith reporting for WPSX from the Farm Show Building in Harrisburg."

Fig. 7-12. Here an educator reports over the telephone. Note the stopwatch, complete outline, and comfortable position. He carefully timed and practiced this report before he called the station.

PLACING THE CALL

After making general arrangements with the station, follow this procedure: Place a call to the radio station. Give your name, and say you have a news report from, as an example, the state fair. Tell the person at the station that you have news about _____ County participants. The person at the station will tell you when everything is ready to tape record your report. Speak up distinctly so the engineer can check the sound level. A countdown (5 . . . 4 . . . 3 . . . 2 . . . 1) before you begin your report may be helpful.

Put it all together and you'll discover that the telephone is an excellent medium for reporting news and special feature material for radio broadcasts.

Merely getting on the air, even at a fairly good time and with a good program, doesn't assure a large, receptive audience. Experience of movie makers and television producers provides sufficient evidence to indicate the importance of program promotion. Entertainers know the value of a good press agent. They spend thousands of dollars to build an image and an audience.

Promotion is particularly essential with educational programs. Viewers or listeners may not be especially motivated to tune in these shows. Somehow you have to make them want to hear what you have to say. There are many ways to do this. Here are a few examples.

Word-of-mouth. You and your co-workers are constantly in touch with the public. There's nothing wrong with plugging your broadcasts at meetings, during office visits, on farm and home visits, at your club, and elsewhere.

Posters. Simple posters that you make yourself can be very effective. For wider distribution you'll want to use more elaborate printed posters. Display them at meetings, in store windows, at grain elevators, or anywhere people you want to reach can see them.

Brochures. Some states make extensive use of brochures to promote radio and television series (see Fig. 7-13). Brochures are particularly useful in pro-

viding your audience with dates, air times, topics, and talent listings. They are a convenient reference.

Newletters. These are ready-made for promoting your broadcasts. People on mailing lists are familiar with what you do and can make up an audience nucleus for any program you start.

Rubber stamps. Costing only a few dollars, these can be used to stamp publications, leaflets, bulletins, and letters that you mail or distribute otherwise. Simple in design, one may contain: "Watch 'Farm-A-Rama,' Channel 3, Saturdays, 1:00 p.m." or "Hear 'Extension Reports,' Mondays–Saturdays, 6:45 a.m., 1340 k.c." Include the station's call letters if you wish.

Special seals. These may be used for special occasions, such as "Twenty Years of Extension Radio on WMAJ." If you have postage-metering machines, use the cancellation printer to call attention to your broadcasts.

Transmittal slips. The government penalty privilege requires that the user's name, title, and signature accompany official mailings. Rather than merely saying, "The enclosed is for your information," or "Enclosed is the information you requested," why not add the schedule of your regular radio and television programs?

Displays or exhibits. Fairs, field days, shows, store windows, bulletin boards, conferences, meetings, conventions, festivals—they all offer opportunities for displays or exhibits, to tell what you're doing in radio and TV.

Special slides. You can use promo slides as "kickers" in slide talks presented in your broadcast area. They could be pictures of you broadcasting, or they could give the air time, day, and station of your broadcasts.

Newspaper promotion. Newspaper promotions are extremely effective, but often difficult to arrange. If you can afford it, advertise on the radio-TV entertainment page and get whatever free promotion the editor will allow. Your chances of getting space are much better if you link the promotion to legitimate news. Note that the topic also will be discussed on radio or TV, telling when and on what station.

Fig. 7-13. Brochures and other printed materials are among the ways you can effectively promote your broadcast services.

Radio and television. Use radio and television to promote radio and television. Spot announcements, 20 or 30 seconds in length, are probably best. Tapes containing authentic sounds and voices can be very persuasive. Lively, colorful slides should accompany the tapes for television. Films or videotapes make the most of the sight, sound, and motion and are probably most effective. Scripts alone aren't as exciting but can be used. In any case, concentrate on the "what," "where," and "when."

Awards. Plaques or certificates presented to a station on special occasions can do much to build strong relations, and they express your appreciation to the station for its support and cooperation (see Fig. 7-14). Get as much news coverage out of award pre-

Fig. 7-14. Personal relationships with station personnel are important. Here two radio station managers receive awards from an educator.

sentations as possible. This shouldn't be difficult. Most stations are awards-minded.

Certificates. Television lends itself especially well to in-depth educational series. A number of states go so far as to have viewers enroll and provide them with home-study materials. At the end of the series, there's usually an examination or some type of evaluation. A splendid way to acknowledge this viewer participation is to present certificates.

Program talent also may be cited for outstanding contributions. Some service clubs present certificates to guest speakers on special occasions; why can't you?

Gifts. Sometimes it's practical to send gifts to members of your audience. Perhaps the gifts represent major subject matter areas to be emphasized. For example, a package of flower seed could tie in with home beautification; a small seedling with reforestation; an apron pattern with beginning clothing construction; or a wildlife and stream map with a conservation project. Such things help sustain long-range interest.

Celebrations. Special occasions, such as program anniversaries, provide a nice change of pace from daily broadcasting. On the anniversary of your program, or of the station's going on the air, break out of the routine. Have a station party—with guests and refreshments. Your broadcast can play a role in the celebration. Promote the celebration, not just on the air but through a variety of media. You'll be surprised how effective a celebration is in maintaining interest in your program, both by viewers and listeners and by the station.

Photo / Art Credits

Bill Ballard, North Carolina State University: Figs. 7-1, 7-3.

Cordell Hatch and *Mike Lynch*, The Pennsylvania State University: Figs. 7-2, 7-5, 7-7, 7-8, 7-10, 7-11, 7-13.

Ron Matason, The Pennsylvania State University: Figs. 7-4, 7-6, 7-9.

Vellie Matthews, North Carolina State University: Figs. 7-12, 7-14.

8 TELEVISION

◀ Communicating via Television ▶

Television has been acclaimed to be one of the most important communication tools available today. Unquestionably, the medium has greatly aided economic and social growth. A better informed public with a more global perspective has resulted.

TELEVISION IS YOUNG AND GROWING

Compared to other media, television is an infant. Although some of the basic inventions which made television possible date back to the late 1800s, television wasn't on public display until the New York World's Fair in 1939.

A certain amount of interest was generated then, but World War II prevented any substantial growth. The affluence after the war and the development of coaxial cable, microwave, and networking promoted rapid growth of "TV," as it became known.

In the 1950s, Ultra High Frequency (UHF) stations entered the telecasting field; they struggled and almost died. Community antenna or cable television (CATV) systems were built, mostly in mountainous states, to improve home reception. A "compatible color" standard was accepted, but few programs or sets were available. And probably most important of all, videotape recording was introduced.

In the 1960s, educational and instructional television made substantial growth. The Public Broadcast-ing Act of 1967 provided funds, as did the Ford Foundation and others. Color telecasting became the standard.

In the 1970s, the idea of a "wired nation" stimulated interest in cable television. With the introduction of low-cost TV cameras and videotape recorders, the door to local origination and public access to CATV channels was opened. Business, industry, education, and government saw the value of non-broadcast television for their internal as well as external communications (see Fig. 8-1).

Electronic news gathering (ENG), with portable, battery-powered equipment, has taken broadcast tel-

Fig. 8-1. A small institutional video production facility.

Prepared by *J. Cordell Hatch,* electronic communications specialist and professor of agricultural communications, The Pennsylvania State University.

evision live or within minutes of where news is made. Time-base correctors make it possible to put these formerly unbroadcastable tapes directly on the air.

In the 1980s, home video recorders, television projectors, pre-recorded videocassettes and disks, cable networks, satellites, computers, teletexts, videogames, and other electronics provide services, program material, and entertainment which was once the almost exclusive domain of motion picture film and regular broadcast television (see Fig. 8-2). Home, personal, and small group use of video will continue to increase in the years ahead.

Fig. 8-2. Low-cost, portable equipment and satellites have opened the door to many new communication opportunities.

WHAT TELEVISION CAN DO

Much of television's success in teaching lies in its unique combination of sight, sound, and motion. This coupling of audio and visual stimuli has proven that it can change human behavior.

Television alters attitudes, conditions thinking, establishes and nurtures cultural standards, and molds public opinion. It educates, informs, entertains. People spend more time with it than with any other medium, averaging over six hours per home per day.

Before discussing television principles and techniques, let's look at what we've learned from research and experience.

- People learn just as much from a televised presentation as from the same or similar material presented in a live, face-to-face situation.

- People of all ages, in all cultures, with all interests learn so many different kinds of subject matter from television that it's difficult to find a topic to which the medium, properly used, can't make a contribution.

- Television helps transcend cultural, political,

geographic, and linguistic barriers and can serve as a unifying, positive force for change, development, and peace.

- If the message can be encoded as sight and sound stimuli, then television can function effectively as the distribution medium.

- More can be said and shown on television in less time than is possible through traditional means.

- Television forces the presenter to prepare. As a result, the clarity and depth of understanding are often better than they would be otherwise. However, television exaggerates lack of preparation.

- Television can swiftly carry news, information, and instruction to large audiences dispersed over wide geographic areas.

- Television often reaches people who can't be or aren't reached in any other way.

- Through television, people become aware of and interested in non-broadcast programs and activities.

- Television gives wide public exposure to communicators and educators. If they use the medium well, they will be recognized and respected for their expertise. Hence, they will gain personal satisfaction, and the rapport they establish with their television clientele will aid in their implementing traditional educational programs.

OTHER ATTRIBUTES OF TELEVISION

A television set can be operated with a minimum of fuss and bother. It can be easily viewed in daylight or darkness. The set has its own built-in P.A. system. Set maintenance isn't a serious problem. There are now receivers in almost all U.S. homes and in many public places.

Television has limitations too. Facilities to originate broadcast programs are costly. Viewers have little control over pacing or content. And usually viewers have no practical feedback mechanism through which they can respond.

Broadcast television has tended to program for mass audiences, and viewer orientation has been one principally of entertainment, not education. However, PTV, CATV, CCTV, ITV, and other forms of television and video production are becoming more readily available and are offering smaller, more specialized audiences a programming and information choice.

TELEVISION IS DEMANDING

Planning, preparing, and presenting television programs that really change viewers' attitudes, increase their knowledge, or stimulate them to action requires much thought and attention. Even the smallest details must be considered to make an ordinary program an outstanding one (see Fig. 8-3).

You, as a communicator, will be able to have good television programs consistently if you possess the following qualities:

● An understanding of television as an effective communication tool.

● A confidence in yourself as an effective television presenter.

● A knowledge of proper television techniques and procedures, coupled with practice and experience in using them.

But these qualities are of little value unless you have something to say. Good television technique or attitude won't compensate for lack of subject matter substance and preparation. Choose solid content and be so familiar with it that you don't have to grope for what to say or do.

As a communicator, you have a number of things going for you. These include:

1. In your professional position, what you do affects people. The kind and degree of effect are closely related to your communication skills.

2. At your fingertips is an arsenal of useful, practical information—most of it based on recent research findings.

Fig. 8-3. Television requires careful planning and preparation as well as attention to the smallest details.

3. In most cases, the television facility and air time are provided to educators at little or no cost. In any case, the cost per person reached usually is quite low.

4. Viewers select you to watch. They get to know you, consider you a friend, and believe in what you say—all important in getting your message across and in changing behavior.

Through television, you can reach people who possibly can't be reached in any other way. And with live television, you can reach extremely large audiences immediately and in a manner not possible by any other communication method.

◄ The Production Crew ►

Few things will boost your television confidence more than having good working relations with the production crew.

Although 10 to 20 people may be involved in getting your show on the air, you will work mainly with the producer, director, floor manager, and host-interviewer, if there is one.

PRODUCER

In some cases this may be you. Generally, the ideas, message development, and production are your responsibility. However, the show on which you appear may have a staff producer assigned. If so, the producer can be a great help in structuring your program.

The staff producer knows the station's facilities, capabilities, and limitations—what can or cannot be done, what will or will not work.

As a rule, producers welcome and appreciate creativity. They want you to come up with a new twist, fresh approach, or different angle to any topic—new or old. They'll find it stimulating; so will your viewers; and so will you.

Sometimes staff producers are also the shows' directors or emcees. Work closely with them; they can be a big help.

DIRECTOR

In the studio the director is completely in charge. The production crew is responsible to this individual who directs all activities during rehearsals and on the air or in a video production (see Fig. 8-4).

Fig. 8-4. The director stages the program and calls all shots during the telecast or taping from the control room.

Generally, the director is your adviser concerning technical matters. When possible, consult by phone before you go to the studio. If your presentation is complicated, you'll certainly want to discuss it ahead of presentation time.

If you have any questions during preparation or rehearsal, don't hesitate to ask the director for help. All props and materials you plan to use (film, slides, photographs, audio tapes, and records) must be checked out and handled by the director.

All details regarding the program's structure, format, and production must be resolved *before* the telecast or taping.

FLOOR MANAGER

The director's right hand in the studio is the floor manager, who, under instructions from the director, sets up the studio prior to a rehearsal or a telecast. During the setup operation, the manager handles flipcards; arranges stands, tables, chairs, and TV monitors; and helps with other preshow activities.

The floor manager stays in the studio during the telecast and is in contact by headphone with the director in the control room at all times. He/she relays instructions from the director to you when you're on the air and gives you start and stop cues. By hand signal, the floor manager will give you other instructions, such as to speed up, to slow down, or to move right or left (see Fig. 8-5). You will see a hand point-

Fig. 8-5. The floor manager works in the studio, conveys directions, and gives time cues.

ing each time the on-the-air camera is about to be changed. You should then look down at your notes and back up at the indicated camera. The camera switch is made while your eyes are lowered.

You won't ordinarily work directly with other members of the production crew, but it helps to be on good terms with them.

◄ Using TV Facilities ►

MICROPHONES

The basic types of microphones used in TV productions are the clip-on or pin-on, the neck or lavaliere, the table, the handheld, and the boom.

Clip-on or pin-on. This is the most popular type of microphone used today in TV and video productions, both in the studio and in the field. It is usually battery-operated and very light, small, and inconspicuous. You can clip or pin it to a tie, lapel, or

Fig. 8-6. The clip-on or pin-on condenser mike has become very popular.

other piece of clothing (see Fig. 8-6). The Sony ECM-50 and the Electro-Voice CO-90 condensor are popular models.

Neck or lavaliere. This type fastens around your neck with a cord or wire holder. It requires little attention. Don't wear pins, necklaces, or tie clasps which may strike against it and cause unwanted noise. Watch out for the cord—keep it out of your way; tuck it under your belt, if you're wearing one. Make the microphone as inconspicuous as possible.

Table. This type of microphone is placed in a stand which sits on a table or desk. Try not to vary your distance from it.

Handheld. Rugged in design, this type is frequently used in field production. It takes the greatest amount of skill to use correctly, especially if used in interview situations involving two or more persons. Hold it steady; don't wave it around like a conductor's baton. Keep your mike arm firmly against your side, not on your chest. Bend your forearm upward to form an angle of about 60 degrees. Avoid moving the microphone from one person to another as each speaks; find a position where everyone is picked up

equally well, and keep the microphone there. Don't be a mike jabber—keep it down and about 12 inches from your mouth. While handholding a mike, don't wear rings, for they will clang against it.

Boom. This microphone is mounted on a long, movable boom operated by a technician. Its chief advantage is that it doesn't show in the television picture. It allows you to move on the set without worrying about cumbersome cables. Of course, the boom operator must anticipate your movements.

CAMERAS

Cameras provide the visual link with your audience. Most stations will furnish two cameras for your studio presentation. Find out how many cameras will be used, and plan your program accordingly. Keep in mind which camera will be doing what as you prepare your script.

Field productions are usually shot using one camera, while large mobile units may have more.

A camera can:

1. Give you intimate eye contact with your audience.

2. Show close-ups of relatively small objects in good detail.

3. Make a small studio or set appear larger.

4. Give the illusion of motion to still pictures by panning on them.

5. Produce an image which can easily be superimposed over another visual or which can be used for matting multiple images.

Fig. 8-7. In or out of the studio, the television camera has limitations which you should know.

A camera cannot:

1. Show the extreme corners and edges of a picture or slide.

2. Show you or your props when either or both are moving about erratically.

3. Effectively show scenes and patterns that are "busy" or cluttered.

4. Anticipate unexpected moves.

5. Show shiny objects.

6. Distinguish between certain shades of color if transmitted or received in black and white.

7. Show black blacks and white whites in the same picture.

8. Adjust quickly to show two or more visuals of widely varying sizes.

When you want eye contact with your viewers, look straight into the lens just as you would look into the eyes of others while you are speaking to them. Don't stare or squint; just look pleasantly "through the lens" to your viewers.

Note the tally light on top. This red light indicates which camera is on. Be aware of it. Also be aware that the lens is very sensitive to quick or extraneous eye movement. The floor manager will give you advance warning when the "on" camera is about to be changed.

VIDEOTAPE RECORDERS

Live television is almost a thing of the past. Practically all television now originates on videotape.

The sound and pictures—30 per second—are laid down as magnetic images on an oxide-coated polyes-

ter tape, 2 inches wide or less. Quadraplex or 2-inch tape has been standard for the broadcast industry. However, the new 1-inch Type-C videotape is now being adopted by many stations and video production houses (see Fig. 8-8). Helican-scan formats use tape on reels, in cassettes, and in cartridges, ranging from ¼ inch to 2 inches wide. Until the introduction of Type-C recorders, helican-scan was used mainly for other purposes—CATV, CCTV, ITV, etc., in schools, businesses, and industries.

A great deal of progress has been made in developing low-cost, reliable helican-scan equipment. A helican-scan recorder costs from $600 to $60,000, compared to $100,000 for a quadraplex machine, and tape cost is one-fifth to one-half as much.

The U-matic videocassette, with its portable recorders and electronic editing decks, makes quality television productions possible outside the broadcast studio (see Fig. 8-9). Portable players which hook up to any TV set extend their utilization greatly. Color cameras costing less than $5,000 also have contributed much to the growth of helican-scan television.

Fig. 8-9. U-matic video editing with time-base correction makes low-cost, high-quality productions possible.

Although editing and re-recording are fairly easy with videotape, most stations approach a taping session as if it were a live broadcast. Studio and equipment time are costly, and retakes are to be avoided if at all possible. Planning and preparation should be the same for taped productions as for live programs. Tape just gives you another chance if you must have it.

REAR PROJECTION SCREEN AND CHROMA KEY

The rear projection screen is a translucent sheet of plastic upon which any visual material using a light source may be projected. It may be installed in a

Fig. 8-8. Computer-controlled 1-inch recorders provide flexibility and quality for telecasts or editing.

studio in a permanent set or as a movable scenery component. It can be found in a wide range of sizes. A large one can form the entire background for an activity.

The chroma key performs electronically the same function as a rear projection screen. It is used by most broadcast stations and networks. News programs make considerable use of it. It gives the impression that the picture is being projected in back of or to the side of the newscaster.

The 35mm slide, 16mm film, and videotape recordings are popular sources for chroma key. Any visual source, including vertical material, may be used. The TV camera can move about on the source—pan, tilt, and zoom—to show detail or to give a sense of motion.

If you plan to use the rear projection screen or chroma key, make advance arrangements with your studio director.

SLIDE-FILM CHAIN

Sometimes called "telecine," this unit is a combination of slide and film projectors coupled to a TV camera. Generally, you may use only 35mm horizontal slides and 16mm film. Some stations may remount your slides between glass. Few stations can handle 8mm film.

OTHER FACILITIES

Your TV production may require an assortment of aids—flannelboards, Velcro boards, chalkboards,

Fig. 8-10. An attractive set and appropriate props enrich the television production.

flipcards, flipcharts, models, real objects, panels, etc. (see Fig. 8-10).

Arrange for only the facilities you can use effectively. A wide variety doesn't necessarily make for a better program. Too many different requirements may actually be a source of trouble. For example, to have slides, and mounted pictures, and film, plus perhaps the real item may be more than is needed, or than can be handled effectively.

A strong presentation often is the simplest one. In any case, the entire TV facility has the primary purpose of helping you convey a message that is relevant, timely, and important, while requiring minimum production cost and viewer effort and motivation to comprehend.

◄ Looking Your Best ►

MAKEUP

In the early days of television, performers relied heavily on makeup to give them their best appearance in front of the camera. With improvements in cameras, tubes, and lighting, makeup is somewhat less important. However, there are various tips which may be helpful.

For Men

1. For those with closely cut hair, delay a haircut until after your telecast. Close haircuts give a scalped appearance.

2. For those who are clean shaven, be freshly shaved. Use a blade razor rather than an electric shaver.

3. For those who are balding, dust the bald spots with face powder to eliminate shiny spots.

For Women

1. For those with blonde or gray eyebrows, darken them with an eyebrow pencil so they won't fade out.

2. Wear lipstick of a light, clear color. A dark blue-red shade tends to be harsh-looking.

3. Have hair well coiffeured. Try to avoid flat or bouffant hairdos, or those which hide the face. Avoid extremes.

4. Make sure other makeup, such as eyeshadow and mascara, is consistent with everyday good grooming. Whatever is natural and comfortable for you is usually natural and comfortable for the viewers. However, if you are light-complexioned, you may need to use heavier makeup to accent facial features.

For Everyone

1. Have well-groomed hands, especially if your hands are to be shown in a close-up. Clean nails clipped to a medium length are best.

2. Make sure hair doesn't look wind-blown.

3. If possible, use the camera and monitor to check yourself before going on the air.

CLOTHING

For Men

1. Stay away from black blacks, which can create harshness, white whites, which can create starkness, and horizontal pin stripes or plaids, which can have a vibrating effect. Thus, a pastel-colored shirt looks better than a white one, as does a solid-colored tie, or one with subdued stripes. Avoid ties with flowered prints or bold, busy designs.

2. Make sure your shoes and socks complement each other and your pants. Those which contrast will divert attention. Wear socks that are long enough that your legs aren't exposed when you are seated.

3. At times when you may be working with soil samples, plants, cattle, or sprays, wear regular work clothes for these demonstrations.

For Women

1. Stay away from black blacks and white whites also.

2. Emphasize vertical lines for a slimming effect. Fitted outfits of firm material usually look best. Check yourself in a mirror at home to see how you will look when you are performing. If it looks good there, it will probably look O.K. on camera.

3. Avoid necklines or hemlines which might cause discomfort or embarrassment if it's necessary to lean forward or to sit in a low chair.

4. Above all, dress appropriately and comfortably for whatever you'll be doing on camera.

For Everyone

1. Wear clothes with conservative colors, yet smartly tailored or styled. These look best on camera. Middle-of-the-road grays, browns, tans, and blues are good. Add a dash of appropriate color in a tie, scarf, sash, etc.—but not too bright.

2. Wear darker shades on light sets; lighter shades on dark sets. Dark-complexioned persons should wear light colors and perform in front of a light background, and vice versa for light-complexioned persons. The center of attention usually is the face; thus, it needs to stand out from the background, not blend into it.

3. Avoid excessive jewelry. It tends to be in poor taste and may reflect light. This includes shiny wristwatches, tie clasps, cuff links, large rings, costume jewelry, bright buttons, or similar objects that flash.

4. Above all, wear something you like, something you feel good in, something that boosts your confidence.

◀ The Television Performance ▶

In order to perform effectively on television and to gain viewer acceptance, individuals must possess some natural attributes for television communications, such as an outgoing personality, charisma, and a pleasing personal appearance. Since television is not like any other medium—with the cameras, cables, lights, people, etc.—a thorough knowledge of what is going on behind the scenes is helpful, as is a knowledge of what the medium can or cannot do well.

Even with the aptitude and knowledge, one usually has to have practice or experience to be really good on television. With experience comes a feeling of confidence, and that confidence is translated by viewers into credibility—probably the most important factor in persuasive communications. An experienced, confident, well-rehearsed performer feels comfortable in the role. This feeling of being relaxed and comfortable is conveyed to the viewers. The uneasy presenter makes viewers uneasy, and uneasy viewers will soon switch channels.

Research shows that "exhibitionist" individuals have better performance scores. In other words, those who like to be in the limelight, the center of attention, do better on television. Aggressive persons with high achievement motivation also do well. On the other hand, persons with high "abasement" and "order" traits do less well. That is, those with a guilt complex and a need for order have trouble adjusting to the unpredictable demands of television. A confident, flexible person copes: a self-conscious, rigid person doesn't.

Regardless of personality traits, a person's ability to perform before the TV cameras improves with age, education, experience, training, practice, and feedback received. A knowledge of subject matter and a mastery of television techniques, coupled with self-confidence and a positive attitude toward TV, usually produce strong television presentations.

Studies also show that evaluative feedback improves performance capabilities. Conscientious presenters will seek comments and critiques from family, friends, station personnel, professionals, and, best of all, from themselves on videotape. Nothing is so instructive as to view one's own performance on videotape. Feedback should be immediate. Self-evaluation should precede peer and expert evaluation. Practice, if you can, on videotape before the actual presentation.

Through study, practice, training, and feedback, any person may improve his/her ability to perform, to present, to teach, to communicate through the medium of television.

PLAYING TO THE CAMERAS

While experience develops skill, it doesn't remove the anxiety or tension of doing television, nor should it. Veterans of many years get "keyed up" before every broadcast. Most perform better when the adrenalin is flowing. The objective is to be neither too "hot" nor too "cold." Try to find a point of "highness" that is right for you, while being comfortable for your viewers. The type of program, format, and subject matter also will affect your intensity level.

Fig. 8-11. When the tally light comes on, look pleasantly "through the lens" to establish eye-to-eye contact with your audience.

If looking at the camera is uncomfortable, then look beyond it—through it—to the camera operator. Direct your presentation to the operator. In studio productions, the director may wish to change from a picture of you on one camera to a picture of you on another camera. If the switch should come without warning, there could be an awkward moment. To avoid confusion, the camera operator or floor manager will give you a cue by holding out an arm and pointing to the other camera.

This signal doesn't mean you should switch your attention directly to the other camera. Instead, when the red tally light goes off, lower your eyes to your notes; then bring them up looking at the other camera. Even better, lower your eyes to your notes when the switch-camera signal is given; then look up at the other camera. The camera switch will have been made while your eyes were lowered. Mastery of camera switching will add to the smoothness of your presentation.

You may wish to practice this at home by using two mirrors placed some distance apart. Think of them as two friends. You've finished talking to one and now wish to relay information to the other.

Use the same mirror setup to practice speaking to one make-believe camera while demonstrating, showing, or displaying to the other. In almost all cases where close-ups are required, you'll be on one camera while the other is in for a tight shot.

With this make-believe studio, go through every word, every motion just as you will when on the air. Don't take any shortcuts. Get your timing down; use a clock or a stopwatch. Rehearse your entire program until you have it just the way you want it. You'll be surprised how this will boost your confidence when you do get before the live cameras.

Remember to speak and display to the imaginary cameras just as you will when on the air.

HANDLING VISUALS

Television cameras tend to make items appear flat. The cameras also exaggerate movements; keep them slow. When handling an object, determine where it will be placed for best visualization. Once this choice has been made, mark the spot and leave the object there. When holding an object for a close-up, maintain its position long enough for the camera to get set up and then take the shot. If the object must be moved for a better viewing angle, rotate it slowly. Don't remove it from the viewing position until your viewers have had time to look at it.

If you have several objects to show, such as in a food demonstration, make arrangements for removing the objects after they have been used. Have a predetermined place to put them. This will eliminate distracting clutter in your presentation area.

TO USE GUESTS OR NOT

If you're an authority on farm, home, and garden matters, don't rely too heavily on guests to tell your story. You may tarnish your image as an educator.

Wouldn't it be better to bone up on a topic, get outside instruction and aid, if necessary, but present the segment yourself rather than have a guest do it? Use a station host-announcer to interview you, if you like. These announcers are trained to handle any crisis on camera, to worry about cues and timing, and to assist you in handling props, etc. This takes a lot of pressure off you.

Some guests are naturals. Others, however, should be chosen carefully. Ask yourself these questions:

● Do the guests really have something to contribute; how will they say it?

● Do they speak often in public; are they nervous; will they freeze up?

● Do they ad-lib freely; are they flexible; must they read or use notes?

● Will they talk too much, jump cues?

● Do they know how to present on TV, handle props, etc.?

Beware of people who have an ax to grind or something to sell. Often they are commercially involved and aren't very objective. Are they concerned with the viewers' needs or their own?

In short, make guest appearances functional and consistent with your aim of instructive objectivity.

IF YOU INTERVIEW, HOW?

An interview is a discussion usually involving two

Fig. 8-12. Guests, properly selected and prepared, can contribute to the success of your program.

or three people. Viewers are considered outsiders and are just watching and listening to you and your guests. Therefore you don't have to look at the camera. The interviewer and interviewee should look at each other and direct questions and responses to each other. As the interviewer, put yourself in the place of your viewers, asking questions that they might ask.

When visuals are used in interviews, your guests should show them to you and explain them, while allowing the viewers at home to see them clearly.

Always prepare your guests by going over the types of questions that will be asked. By explaining the purpose of the interview, you will help your guests become more a part of the program. This also helps them relax.

Direct your questions in such a manner that your guests have a chance to respond and talk. Don't ask questions starting with "Did you . . . ?" "Have you . . . ?" and the like. This invites direct "yes" or "no" answers. Ask "why," "how," "who," "what," "where," or "when"—especially "why" and "how." Usually questions of this type trigger the most interesting replies.

TV TEAM APPROACH

Rarely do inexperienced television performers handle visuals and manipulate objects on the air properly. Generally, movements are too fast, advance cues to the director and camera operator aren't given, and the continuity of thought is broken.

The team approach eliminates, or at least reduces, these problems. One person is on camera and devotes full attention to telling the story or describing the action. On another camera the teammate handles all models or objects, manipulates, demonstrates, letters, or points at specific areas to focus the viewers' attention there.

In short, the teammate does everything possible to relieve the speaker of all distracting tasks. This person may or may not be on camera. That depends on the activity. If the shots are all close-ups, then only the hands will show. There will be times however, as when handling livestock or pruning trees, that this teammate will come into full camera view. Then the teammate becomes a recognizable part of the program and should be identified on the air as such.

The team approach usually results in a *smoother, more professional-appearing program* than if the speaker tries to do everything alone. Newcomers usually do a better job when they have only one phase of the telecast to worry about.

Fig. 8-13. Sometimes two persons are needed to do a presentation well.

◀ Television Visuals ▶

Much of television's effectiveness lies in its visual content. Is there a pictorial impact on the viewers? Don't get caught producing a televised radio program. Another caution: Poor visuals may actually distract from rather than support what's being said.

In television, the visual message and audio message are assimilated as one. The viewers' final interpretation includes both audio and video.

Fig. 8-14. Select visuals that will help tell your story and give a pictorial impact on the viewers.

Good visuals attract attention, arouse interest, add emphasis, give variety, explain points, and clarify information. Visuals also:

1. Help to overcome language difficulties—ambiguities.

2. Show size, shape, color, texture, similarities and differences, change, sequence, and relationships.

3. Help to bridge time (bring summer into winter).

4. Help to bridge space (bring a barn into a TV studio).

Simplicity has many virtues, particularly in television. Your most effective visual generally is a simple one with only one or two major points. The complex, with many details, is often confusing; one point competes with another for viewer interest and attention. Help focus the viewers' concentration by having them see only one mental picture at a time. The complicated visual, in addition, is likely to be more expensive and time-consuming to prepare. This is especially true with artwork—illustrations, charts, graphs, forms, figures.

A VISUAL ISN'T A VISUAL UNLESS IT'S VISIBLE

Visibility depends on:

1. Brightness contrast—that is, the contrast between the light and dark areas of the visual.

2. Color contrast—which is usually combined with brightness contrast to aid visibility.

3. Size.

4. The viewers' eyesight.

In practice, you may want to incorporate *safety factors* for those viewers whose eyesight isn't the best. Keep in mind the *comfort factor* too. If viewers have to strain too hard to see, they will give up. In addition, a large image is generally more impressive than a small one.

Make TV visuals like billboards—so the viewers can get the message at a glance.

THE REAL THING

Usually the real thing is your best visual. If you're explaining how to barbecue a chicken, barbecue one. Nothing holds interest like the real thing, whether it be yourself or some object.

Since use of the real thing isn't always practical or possible in the TV studio, you may have to rely on something else to convey your visual message. There are dozens of different methods and devices you can use. If one of the following doesn't fit, a little creative thinking is likely to turn up a solution.

TELEVISION CARDS

Sheets of 22-inch × 28-inch posterboard are available in many colors at most stationery and art stores. One piece can be cut into four 11-inch × 14-inch television cards. Gray, light blue, and light green are useful colors; they give appropriate contrast with black or dark color lettering or drawing.

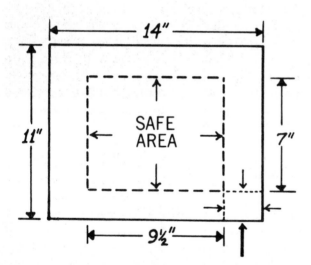

Fig. 8-15. Lettering and essential information should be positioned on the TV card in the center safe area. Margins are easily cropped off by TV sets and cameras.

For maximum legibility:

1. Use no more than five lines of lettering per card.

2. Seldom use more than 15 letters and spaces per line.

3. Use block-style letters.

4. Make letter height ¾ inch to 1 inch, or higher.

5. Make letter width about ½ inch.

6. Make stroke width about ⅛ inch.

7. Separate letters and numbers by one stroke width, ⅛ inch.

8. Use only the 7-inch × 9½-inch safe-copy area in the center of the card, leaving 2-inch top and bottom margins and 2¼-inch side margins (see Fig. 8-15).

9. Remember that the TV screen's proportions are 3 high × 4 wide.

SLIDES (35mm)

Color slides are widely used on TV programs. They're readily available, easy to take, and easy to use. Keep the following in mind:

● Only about two-thirds of a slide's picture area will be seen by the home viewers. That, of course, is the center area (see Fig. 8-16).

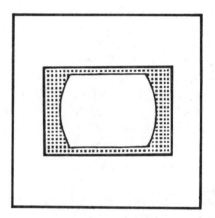

Fig. 8-16. Position your 2-inch × 2-inch slides over this mounting. Only the center portion will be seen on home TV sets.

● Vertical slides aren't usable in the slide chain. They may, however, be chroma keyed or projected on a rear projection screen for direct camera pickup.

● Slightly overexposed slides are better than underexposed ones. If the slides are taken expressly for television, open the shutter an extra one-half f-stop. This will give you thinner, brighter projecting slides.

● The outer edges of a slide won't be seen by the home viewers. Allow for this in taking pictures.

● On black-and-white sets, the colors will be converted to shades of gray. Light the subject properly and provide a background for suitable contrast.

Four, five, or more slides per minute of talk can be used. Also, slides are easily intermixed with other types of visual material, including on-camera pictures of the talent.

PHOTOGRAPHS

The same general principles given for slides apply, plus the following:

● Glue, cement, or tape photographs securely to mounting cards.

● Generally, mount only one picture per card. If you do mount more, leave enough space on each side of a picture so that the edges of other pictures don't show. Remember the shape of the TV screen—3 high × 4 wide.

● If you are going to flip cards on the air, make sure all pictures are the same size and mount all of them in the same position on all cards. The TV camera shouldn't have to move for on-the-air flips.

● Eliminate distracting picture backgrounds by cutting out in silhouette the interest area, then mounting the cut-out on a contrasting card. This will give a nice effect and will allow you wide latitude of picture sources—magazines, catalogues, publications, leaflets, etc. Remember copyrighted materials may require written permission and a credit line. Be sure you're legal before using them!

● Prepare montages. They are effective and fairly simple to make. They will provide a good background for title and credit supers, and they may be suitable for camera panning from one interest area to another.

MODELS AND EXHIBITS

Models sometimes are easier to work with and can show more detail than real objects. Often they can be assembled or disassembled on camera. For example, it's easier and more effective to show rings, pistons, valves, etc., in a model than it would be to drive a tractor into the studio to try to show the engine and explain how it works.

LIVE GRAPHICS

On camera, you may draw on cards, newsprint pads, chalkboards, or acetates on an overhead projector. You should keep the writing or drawings simple and do them with heavy lines. However, you

may outline the material ahead of time with a sharp, hard lead pencil.

You can prepare strip, or tease, charts beforehand. You cover the information with a strip of paper and reveal it on camera at the appropriate time.

Live graphics are especially effective in building sequences and in creating suspense. Elements can be displayed in any sequence to enhance the final message.

FILM—MOTION PICTURES

Film preceded videotape as a production medium for television. The equipment is relatively portable, economical, reliable, and easy to operate. However, most on-location shows now originate on videotape.

Fig. 8-17. Motion picture film has lost ground to videotape but still has a place in television productions.

Even so, film, either silent or sound, still has a place in taped or live telecasts. Programs completely on film are expensive and time-consuming to produce, but carefully selected portions or concepts on film can enhance interest and understanding.

Segments of motion pictures may be included in your telecast. They're available from many libraries; 16mm is standard for television. Most films produced today are cleared for television; nevertheless, you'd better check.

VIDEOTAPE

A variety of videotape equipment is available which will allow field or studio production of segments or full programs. Even helical-scan tapes can be aired if processed through a time-base corrector. Videotape production and utilization is similar to

Fig. 8-18. Videocassettes make production, editing, and playback easy and economical.

that of film, except that one is electronic, the other photographic.

CHARTS, MAPS, GRAPHS

Limit a graph to two comparisons. Large letters and few words are necessary to maintain simplicity. Materials should be well centered. Use contrasting colors and patterns on maps and charts for emphasis and comparison.

CUE CARDS

Do your own hand-lettered cue cards—if you feel you need them. Cue cards with an outline of your facts may well aid you in making a better appearance and help keep your presentation precise and well organized. They can be held for you near the camera lens. However, this is sometimes 10 to 12 feet away, so be sure you can read your lettering at that distance. Electronic teleprompters are available in most modern studios (see Fig. 8-19).

Fig. 8-19. TV teleprompters allow you to read copy and look directly into the "taking" lens at the same time, thus maintaining good viewer eye contact.

PUBLISHED ART

Lithograph materials, art posters, and pictures designed for uses other than for television may be used without further processing. It's possible to take portions of this material and combine or rearrange them to suit your needs, but don't violate the copyright law.

OVERHEAD PROJECTUALS

These are 8 × 10 transparencies—called cells, or acetates. While showing an image on the screen, you can write or draw on a transparency. This permits a gradual buildup to a final message. Transparencies can be superimposed—one laid over the other (overlays). Overheads have limited television use.

CHALKBOARD

Use soft chalk, pastel colors, and large letters. Clean erasure is a problem. Chalkboards aren't widely used.

PAPER PAD ON EASEL

This is a substitute for the chalkboard. Horizontal pads are preferred to vertical ones. You may do the lettering before air time on a gray or off-white paper. Your local newspaper may have roll ends of newsprint paper which can be cut to desired size and stapled to plywood or paper board. Again, do lettering and artwork carefully.

FLANNELBOARD AND VELCRO BOARD

Both flannelboard and Velcro board are useful and versatile. They work well in building sequences and showing relationships, and they make for an active presentation. You can move symbols. You can use color.

Make your own flannelboard by stapling ordinary flannel (gray or pastel) to board or plywood. Be sure the flannel is smooth over the board. Visuals stick to the flannel if they are backed with sandpaper, felt, or other such materials. Tilt the flannelboard slightly backward so the visuals adhere better. Velcro or hook-and-loop boards will support heavy, three-dimensional objects.

REAL THINGS: RE-EMPHASIZED

Since television reflects reality, it follows that some of the best visuals are real objects. If you're going to be talking about a flower, show one; about sewing a dress, sew one; about pruning a shrub, prune one (see Fig. 8-20).

If you're going to be talking about a pet or a sewing machine or something else that is physically possible to get into the studio without undue bother, then make it your visual. It'll give punch and impact to your program. Obviously, if you're going to be talking about the operating principles of a two-way centripetal pump, a cross-section drawing or a cutaway model of its parts would be much more effective than showing the outside of the pump. And, there's no way you can bring an irrigation system into the studio.

CLEARANCES

Music, films, pictures, and other copyrighted materials should be cleared with the station before you use them on the air.

Most stations subscribe to BMI and ASCAP music representatives. This allows them to perform or play recorded music on the air.

Most films produced today are "cleared for television." This means that the people in a film and its producers consent to its being shown on television.

Fig. 8-20. Real objects usually make the best visuals. Do something to or with them.

Film catalogues will indicate whether or not a film has been cleared for television.

Pictures and other protected material may sometimes be used on the air. But it's advisable to check all such material with the station management and get written permission for its use from the rights holders.

◀ Six D's of Creativity ▶

Creativity is as old as humankind itself; yet, no one has succeeded in isolating its component parts.

Some argue that creativity is inborn. Others contend it's learned. Still others say it's a product of the times and the environment. Best indications are that it's a combination of all three.

Scientists and educators seem to know more about what stifles creativity than what stimulates it. The three major stiflers of creativity are *tradition, fear of failure,* and *habit.*

Rather than explore these negative forces, let's think of ways to achieve greater creativity. Remember that creativity—that power to be original, different, ingenious, productive—can have results ranging from utter failure to complete success.

Creativity, to succeed, must be accepted. This acceptance may not come for years in art, but it must be immediate in television. The fate of the show, program, or series depends on it.

What can you do to help assure television success? While a number of things can be done, success—even with the top professionals—can't always be

guaranteed. However, shows that exhibit a high degree of creativity generally have a better chance to succeed. This is as true with educational TV programs as with entertainment programs, perhaps more so.

Since creativity doesn't come easily, and there's no secret formula for success, what can you do that might lead to greater television creativity and success?

Let's assume you already have clearly in mind what you want the television audience to **know, feel,** and **do.** This is prerequisite and basic to any communication.

Next is the problem of how to present the subject matter—your **message**—in a creative fashion. Perhaps you'll find these six steps useful.

1. **Dream.** Think of different ways to present your ideas, your program. Uncap your mind to all possible approaches and techniques. Let your imagination search, probe, ponder. Don't be bound by tradition and habit at this stage. You

don't have to do your program the way you did it last time or the way your co-workers do theirs. Break the stereotype chain.

2. **Discuss.** Talk over your ideas with co-workers; try them out on members of your own family. See how the television director or producer reacts to them. Consider all thoughts and suggestions.

3. **Dig.** Study, re-examine. Fragment your ideas, take them apart, look at them, and then put them together again. Search for a better way to arrange the pieces. Do you have or can you get all the equipment, materials, and people you need to carry out your ideas—to do your program?

4. **Decide.** How can you use these resources to best advantage? Decide what you will show and tell and what others must do. Bring your ideas into focus.

5. **Design.** Plan what, when, where, and how events will happen during your program. Make a presentation blueprint. This is your script, or rundown sheet. Try to anticipate trouble spots. Know what alternate approaches you can take if problems occur during actual presentation.

6. **Do.** Carry out the plan to the best of your ability.

SELECTING A TOPIC

Although your creative genius plays its major role after the topic has been selected, a little organized thinking at this point is helpful also.

Your guiding principle should be: *Select a topic that is **needed**, not merely one that is **easy**.*

How do you know what's needed? Let problems clue you. What are people asking about? What do they want help with? What are the office and telephone questions? What are people talking about? What's happening—events, activities? Check your calendar or datebook.

What do your viewers really need to **know, feel, do?** What are their needs and goals? How can you help fulfill those needs or reach those goals? Your success and personal satisfaction are determined mainly by how well you do this. If your viewers aren't helped, you haven't accomplished your mission.

Other factors include:

● Will the topic appeal to urban as well as rural viewers?

● Is the topic suitable for television?

● Would some other medium or method be more efficient?

● Does it capitalize on TV's sight, sound, motion?

● Are you qualified to present the topic?

● Can you get all the facts and information you need?

● When is the best time for the presentation?

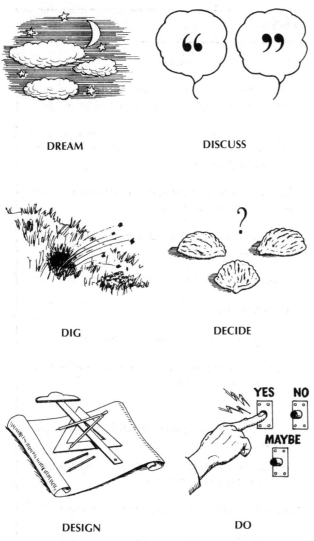

DREAM DISCUSS

DIG DECIDE

DESIGN DO

Fig. 8-21. Employ the six D's of creativity in your television productions.

Television is a complex medium. It requires much equipment and a large staff of technicians. To coordinate all this in getting a show on the air requires a blueprint for action—an outline, a rundown sheet, or a script.

The paper preparation needed depends on the type of show you have in mind and, to a large extent, on what the director prefers.

Shows with many visuals and much action require more thorough preparation than simple, straightforward presentations. Regardless of the type of show, the major components of a TV program are video and audio, what you'll show and say, or better still—what your viewers will see and hear. Even though most shows may be ad-libbed or semi-scripted, they need careful preparation.

Work it out. The particular shot or sequence you plan on paper may not be mechanically or technically possible. Discuss it with the director to see if a better way may be found. Be flexible—compromise if you must—but don't forfeit a good idea just for the sake of making things easy.

Sample TV Rundown Sheet	
VIDEO	**AUDIO**
Open—Close-up (CU): Live crickets in jar.	Smith: "These little critters are a household nuisance. They damage fabrics and sometimes crops."
Medium Shot (MS): Smith with field specimen of cricket damage.	Announcer: "It's time for Farm and Home News—and for today's report, here is your County Extension Agent, John Smith."
Close-up of damage to plant (MS): Smith	Smith: "Let's look at some of the ways we can control these insects—crickets, etc."
(CU): Weatherstripping of board	"Proper weatherstripping keeps insects outside the house."
(CU): Piece of pipe through large hole in board	"Space around pipe should be sealed."
(MS) to (CU): Demonstration of how to seal pipe	Demonstrate method.
(CU): Zipper bag	"Proper storage of clothes is important."
(MS) of three types of insecticides	"Protection is fine, but we want to do away with these insects altogether."
(CU) of: 1. Can of dusting powder 2. Liquid preparation 3. Bag of wettable powder	"Here are three types of insecticides that can do the job." Explains and shows how to apply insecticides, with emphasis on safe use, etc.
(CU): Bulletin Card: Write to County Agent, Courthouse, etc. (MS): Announcer	"For further information, write for this bulletin." "That's our farm and home feature for today. See you next week."

You must know what you want your viewers to **know, feel,** and **do.**

The opening should get your viewers' attention immediately. Open with your ace—this may be an unusual fact, visual, or movement; a challenging, thought-provoking statement; something to stimulate curiosity; or a rather direct statement of why your topic is important or timely. For example, how will your message affect the viewers' income, health, well-being, happiness?

Gather facts to support your major point. Use whatever evidence you can find to make your message acceptable to your viewers—they're the people you have to sell or convince.

A cushion is needed so that you can fill whatever time is given you. It's planned into your rundown sheet, or outline, so that you can expand or condense as the time requires. Be able to cut material without weakening your main point or cushion, without hurting interest or becoming trivial.

The summary will probably do as much to get your message across as anything. It should re-emphasize not merely repeat the important points, and if ap-

Fig. 8-22. Television is action oriented; do something, show things.

propriate, call attention to sources of additional information, such as publications and bulletins. *Do not* rehash the whole show, overdo the thanks to guests, or forget to plug your next show.

◀ Television Do's and Don'ts ▶

YOUR PREPARATION	
Do—	*Don't—*
● Regard television as a unique medium of communication.	● Assume radio and television are alike.
● Make friends with the station personnel.	● Forget other individuals are involved in putting your program on the air.
● Be real, sincere, true to life.	● Be pretentious, misleading, false.
● Develop a good reputation; make viewers want to watch your program to get your message.	● Cause the viewers to say, "There's that character again. I've seen enough of this program."
● Prepare carefully . . . understandingly.	● Assume anything or omit any details.
● Check and double-check all props.	● Ignore proper placements of each prop.
● Keep introductions and announcements short.	● Give the life history of your guests before interviewing them.

(Continued)

YOUR PREPARATION (CONTINUED)

Do—

- Time yourself and your presentation carefully. Practice your closing remarks and time them.

- Know your audience.

- Use the type of program that will best show what you want. Involve your viewers.

- Think visually.

- Use live animals or real objects whenever possible.

Don't—

- Guess at how long each segment of the program will be, and present each haphazardly.

- Ignore audience interests, needs.

- Use a lecture when an interview or a demonstration might be more effective for your message.

- Ignore visual impact.

- Use pictures, unless it's impossible to use the real thing.

YOUR CAMERA PRESENCE

Do—

- Acknowledge any accidents and failures and then proceed with the presentation.

- Be at ease.

- Look at the camera with the tally light.

- Avoid mannerisms. Make all movements slow and deliberate.

- Ignore distractions (cameras and studio crew movements during presentation).

- Be aware of camera change.

- Stay in the position agreed upon during rehearsal.

- Watch the floor manager for time and camera cues.

- Refer to the picture on the screen as "This _____."

- Establish the picture; then explain it.

Don't—

- Let accidents upset you to the point you lose continuity of the program.

- Be tense.

- Watch the monitor.

- Gesture at arm's length or make quick movements.

- Let studio activity upset you.

- Assume one camera is on you all the time.

- Move to an unrehearsed position where the camera can't follow or the light is bad.

- Let viewers see you acknowledge cues.

- Say, "On the monitor you see _____."

- Explain the obvious.

YOUR TALK

Do—

- Talk slowly, simply, clearly, logically, candidly.

- Talk conversationally, as if to camera operator.

Don't—

- Talk so fast no one can understand you.

- Use difficult words, complex sentences.

(Continued)

YOUR TALK (CONTINUED)

Do—

- Talk convincingly—persuasively.

- Talk primarily to the viewer.

- Remember you must compete for attention.

- Satisfy the eye and the ear.

- Combine words with pleasing action. Speak in thoughts, not words.

- Show enthusiasm.

Don't—

- Assume you have points when there are none.

- Hedge or use double-talk. Preach or apologize.

- Think you are the viewer's only choice.

- Talk as if you were in a lecture or convention hall.

- Be blasé about your subject.

- Act bored or half-hearted.

YOUR ATTITUDE

Do—

- Be yourself.

- Be alert, friendly, sincere, personal.

- Have purpose and conviction.

- Bring your viewers into your confidence.

- Think of one imaginary person—a stereotype of the audience you're trying to reach.

Don't—

- Be superficial, or have a know-it-all attitude.

- Be starchy, stilted, unnatural.

- Be distant or impersonal.

- Talk down to your viewers.

- Think of your audience as a crowd.

YOUR PERSONAL APPEARANCE

Do—

- Stay with basically conservative colors and styles of clothing. Avoid extremes.

- Sit up straight; lean slightly forward.

- Dress modestly; be neat.

Don't—

- Wear white, black, bright colors, pin stripes or plaids, bold patterns.

- Slouch and have poor posture.

- Wear excessive jewelry; look windblown.

TELEVISION INTERVIEWS

Do—

- Welcome your guests.

- Go over questions with your guests before air time.

- Explain operations in the studio before air time.

Don't—

- Neglect your guests so they become uncomfortable.

- Ask questions your guests aren't prepared for.

- Forget to show guests around the studio to acquaint them with its facilities.

(Continued)

Do—	Don't—
● Pick key questions to make the best use of your guests and air time.	● Ask meaningless or impertinent questions.
● Make cues and signals known to your guests.	● Assume your guests know TV terms and cues.
● Follow conversation—make sure questions follow logically.	● Stick to your list of questions, in order, regardless of what a guest has said.
● Be interested in your guests and what they have to say.	● Overemphasize your own contribution to the program.
● If your guests are experts in a field, treat them as such.	● Try to impress your guests and/or viewers with your knowledge of the subject being discussed.
● Put yourself in the place of your viewers—ask questions they might ask if they were there.	● Ask questions of interest only to you—personally.
● Ask your guests to clarify any points that seem confusing.	● Feel you can't interrupt.
● Have eye contact with your guests and viewers (camera).	● Look at the camera when you are speaking to your guests.

◄ Evaluating Your Shows ►

Evaluation isn't always pleasant. Perhaps this is why it's so frequently neglected or overlooked.

On television, you, the presenter, are constantly being evaluated by the viewing public. This is good; it helps keep you on your toes.

Repetition of bad practices and techniques becomes just as much habit as repetition of good ones. Therefore, it's important for you to evaluate your TV performances occasionally so that you can determine your strengths and weaknesses.

AUDIENCE RESPONSE

How do you test audience response? One way is to analyze the number of cards and letters or publication requests received. This will measure response, but it isn't an altogether reliable method for determining "who" or "how many" are watching. Requests for publications may depend more on what was offered and how it was offered than on audience size and makeup. Many people just don't respond to write-in appeals. Don't mark them off, however; they may be your most faithful followers. Poor response also may be due to unclear instructions.

SURVEYS

Surveys are your most accurate measure of who and how many are watching. Stations take frequent audience polls. Perhaps they can tell you what you need to know. If not, devise a survey of your own. Ask only questions that will give you useful information. Here are three types of surveys:

1. *Mail questionnaires* for this purpose haven't been very satisfactory. Returns usually are low.

2. *Telephone interviews* are widely used and are probably the most practical. They don't cost very much and are fairly easy to conduct.

   Here's how to go about such a survey: First, develop the questions. Next, determine the size sample you need by area or telephone exchange. The names may be chosen at random or according to some system, such as selecting every twentieth or thirtieth name. Finally, assign the names to your crew of callers, with instructions on how to conduct the interviews.

3. *In-person interviews* are reliable, but costly. The high expense usually rules out this method.

BEHAVIOR CHANGE

Behavior change is very difficult to measure. In the first place, at least two tests are required—one before information exposure and one after. At each checking, you have to determine what the viewers really **know, feel,** and **do.** Is there a change? Is it significant? Is it the result of your television work—in part or whole, or is it something else?

EASY LISTENING FORMULA (ELF)

Can your TV program pass the listenability test? The following formula can be applied to the audio portion of your show. It's easier to use than most readability formulas.

In each sentence to be checked for listenability, count only those syllables above one per word. The average per sentence should be less than 12.

For example, the first sentence in the preceding paragraph has an ELF score of 10: "Sentence,"

"only," and "above" score 1 each; "syllables" scores 2; and "listenability," 5. The second sentence has an ELF score of 3. Only "average" and "sentence" have more than one syllable per word. Thus, the average for the two sentences is 6.5.

This formula may seem rigidly prescriptive. In operation, it needn't be. Writers are free to establish their own level of listenability based on this guide. However, keep in mind that the most highly rated network television news writers use a style that averages less than 12. Walter Cronkite used from 9.6 to 11.9.

This doesn't mean that all sentences should have no more than 12 syllables above one per word. A sentence with 20 or more may be perfectly clear. It depends on the structure of the statement and the nature of the concepts expressed. For variety and ease of pacing, follow a long sentence with a short one.

The Easy Listening Formula doesn't discourage the long sentence, provided the sentence contains short words, which usually means simple words. Nor does it discourage the use of long, complex words, provided a complex word is nested in a short sentence.

But, it does discourage long, complex sentences that confuse listeners, who lack the readers' opportunity to review, digest, and mull over a sentence. In short, avoid the long sentence with several concepts, subordinate clauses, and prepositional phrases.

TELEVISION EVALUATION SHEET		
	(Circle Choice)	
Organization		
1. Captured viewers' attention immediately	6 5 4 3 2 1	Had no attention-getter
2. Aroused interest in subject	6 5 4 3 2 1	Bored viewers
3. Had mass audience appeal	6 5 4 3 2 1	Interested few viewers
4. Had beginning, middle, end	6 5 4 3 2 1	Rambled, fell apart
5. Points simple, direct, easily remembered	6 5 4 3 2 1	Confusing, evasive
6. Content developed logically	6 5 4 3 2 1	Illogical
7. Enough points for available air time	6 5 4 3 2 1	Too few ( ); too many ( )
8. Content closely related to viewers' needs	6 5 4 3 2 1	Irrelevant
9. Interest held throughout program	6 5 4 3 2 1	Interest lost

(Continued)

(Circle Choice)

Content

10. Timely   6 5 4 3 2 1   Outdated

11. Accurate   6 5 4 3 2 1   Inaccurate

12. Based on scientific findings   6 5 4 3 2 1   Opinions, guesses

13. Repetitive emphasis sufficient   6 5 4 3 2 1   Omitted (   ); belabored (   )

14. Important points summarized   6 5 4 3 2 1   Ended abruptly

Message

15. Objectives clear   6 5 4 3 2 1   Ambiguous

16. Motivated change in behavior   6 5 4 3 2 1   Failed to motivate

17. Further interest in topic stimulated   6 5 4 3 2 1   Discouraged

18. Entertained as well as educated   6 5 4 3 2 1   Wasn't fun to watch

19. Helped meet group goals   6 5 4 3 2 1   Wasn't related

Visuals

20. Adequate number used   6 5 4 3 2 1   Too few (   ); too many (   )

21. Clearly visible, perceivable   6 5 4 3 2 1   Unclear

22. Neatly prepared   6 5 4 3 2 1   Unattractive

23. Same proportions as TV screen   6 5 4 3 2 1   Wrong proportions

24. Coordinated with what was said   6 5 4 3 2 1   Unrelated to message

25. Helped clarify ideas, points, message   6 5 4 3 2 1   Diverted attention

26. Studio props and set flattering   6 5 4 3 2 1   Detracting

27. Overall visualization effective   6 5 4 3 2 1   Ineffective

Audio

28. Audio used well   6 5 4 3 2 1   Poorly

Presenter

29. Confident, poised   6 5 4 3 2 1   Uncertain, ill at ease

30. Warm, pleasant, friendly   6 5 4 3 2 1   Cold, harsh, distant

31. Articulate, enunciated well   6 5 4 3 2 1   Inarticulate, enunciated poorly

(Continued)

TELEVISION EVALUATION SHEET (CONTINUED)

Presenter   *(Circle Choice)*

32. Good breath control   6 5 4 3 2 1   Poor breath control

33. Good voice inflection, emphasis   6 5 4 3 2 1   Artificial inflection, poor emphasis

34. Appropriate gestures   6 5 4 3 2 1   No gestures (   ); inappropriate ones (   )

35. Interested, sincere   6 5 4 3 2 1   Unconcerned, insincere

36. Good sense of humor   6 5 4 3 2 1   Humor lacking

37. Stimulating, enthused   6 5 4 3 2 1   Apathetic, boring

38. Creative, imaginative   6 5 4 3 2 1   Non-creative, non-imaginative

39. Flexible   6 5 4 3 2 1   Rigid

40. Informed, knew subject   6 5 4 3 2 1   Lacked knowledge of subject

41. Vocabulary suitable   6 5 4 3 2 1   Unsuitable

42. Neat in appearance   6 5 4 3 2 1   Untidy

43. Dressed appropriately   6 5 4 3 2 1   Inappropriately

44. Spoke to TV camera as if it were viewer 6 5 4 3 2 1   Lacked camera-viewer rapport

45. Physical movements slow, smooth   6 5 4 3 2 1   Fast, jerky

46. Objects and visuals displayed properly   6 5 4 3 2 1   Difficult for camera to pick up

Presentation

47. Overall effectiveness: excellent   6 5 4 3 2 1   Poor

What did you like most about the presentation?   Why? _____

_____

_____

What did you dislike most about the presentation?   Why? _____

_____

_____

Adapted from The Pennsylvania State University Television Evaluation Sheet.

The prospects are bright for expanded and more meaningful use of television in extension education. Extension workers have used television to teach specific audiences, with the viewers enrolled in these televised classes receiving study materials. You may find the following guides to be effective.

● Limit the televised subject matter and printed support materials to a particular problem area and design and present it with a specific audience in mind.

● Supplement TV presentations with materials specifically designed for this purpose. Materials designed for other purposes often don't accomplish the desired TV instructional objective. In fact, non-integrated materials may actually deter potential clientele from viewing the TV presentations.

Fig. 8-23. Video disks and large screen television offer many instructional possibilities.

● Restructure old or familiar materials to coincide with the TV programs. Emphasize how principles and concepts may be applied. Integrate all elements into the total learning process. Individuals want and need help in applying knowledge. *Knowing and knowing how to apply knowledge* aren't the same.

● Do programming on a regular basis or on a well-publicized, special, sequential block basis, if it's to be most effective. Viewer pre-enrollment and introductory materials are particularly needed for the short-run instructional series. Publicity and enrollment also help motivate viewers to make more effective use of integrated support materials.

● A special series presented daily for a short period of time may have some advantage over the same programs presented on a weekly basis. Viewing interest of adults appears easier to sustain for the shorter span of time.

● When preparing and presenting materials related to the farm enterprise, the TV presenter should consider both men and women as the target audience.

● Participants enthusiastically support group TV viewing. Viewers in groups seem to receive (at least psychologically) more help from the telecasts and from materials than do viewers watching individually at home.

● A viewing group benefits from the presence of a qualified discussion leader (one who knows group-learning processes and the subject being taught). This leader conditions the group for the presentation of TV stimulus material, sees that viewing takes place under the best possible conditions, and after the TV presentations, answers questions and coordinates discussion and problem solving. The leader is also active in program evaluation and follow-up.

● No special facilities are necessary for group TV viewing to be successful. A large living room, a TV dealer's showroom, a cocktail lounge, a restaurant, a church, a motel or hotel, a local factory, a school, or a university is an acceptable site.

● Arrangements for group meetings take only slightly more time or effort than arrangements for traditional meetings. Often it's easier to set up and operate a TV set than an assortment of other audio-visual devices.

- TV presentations can make effective use of specialists and guest instructors in reaching large clientele groups. There can be considerable savings in both time and travel costs.

- The majority of extension TV program viewers don't participate in extension in other ways. For example, homemakers who have babysitting problems can't attend extension meetings but can and do watch television.

- Television represents one means of distributing special materials to more than one county, thereby permitting county personnel to specialize in certain subject matter areas and to share this knowledge outside their own counties.

Thus, there is ample evidence of television's capacity to carry organized learning experiences to viewers in structured situations. Bold forward steps need to be taken.

Photo / Art Credits

Bill Ballard, North Carolina State University: Fig. 8-21.

Ron Matason, The Pennsylvania State University: Figs. 8-2, 8-8, 8-9, 8-19, 8-23.

Vellie Matthews, North Carolina State University: Figs. 8-6, 8-12.

Mississippi State University: Fig. 8-4.

Third Edition: Figs. 8-1, 8-5, 8-11, 8-13, 8-14, 8-15, 8-16, 8-18, 8-20, 8-22.

Steve Williams, The Pennsylvania State University: Figs. 8-3, 8-7, 8-17.

WPSX-TV, The Pennsylvania State University: Fig. 8-10.

9 BASIC PHOTOGRAPHY

Photography makes time stand still. It captures activities you wish to remember, programs you need to document, and people you never want to forget.

This chapter is written for the person who wishes to communicate with photos. It will cover (1) selecting a camera that's just right for you, (2) adjusting the camera, (3) making your pictures more enjoyable and informative, (4) employing special photographic techniques, (5) buying and using camera accessories, and (6) creating a slide series and a filing system.

A camera is an important tool for communicators. But don't forget that the photographer, not the camera, is the most important tool for taking good pictures.

With creativity and determination, you can take exciting photographs by using simple and relatively inexpensive equipment. Properly selected equipment will allow you to concentrate on taking pictures that convey a clear message instead of fumbling with camera mechanics.

◀ Selecting the Camera ▶

No camera is perfect for every purpose. You must decide what advantages—such as simplicity—you are willing to trade for others—such as versatility.

Your choice should depend on the kinds of pictures you will take and how often you will be using the camera (see Fig. 9-1). If you need enlargements for public displays, you should select a camera different from one you would use to take snapshots for an office report.

Also, consider how many people will use the camera and how much they know about cameras. If the camera will be used by several people who have little photographic background, select a simple and durable model. Extra features add flexibility, but only if you know how to use them.

Fig. 9-1. You should select a camera and accessories based on how you plan to use the pictures you take.

Prepared by **Jeanne Gleason,** associate agricultural editor, New Mexico State University, and **Don Breneman,** audio-visual specialist, University of Minnesota.

CAMERA CLASSIFICATIONS

Cameras are classified by format, which is the size and shape of image produced on the film (see Fig. 9-2). All cameras will take prints and slides, depending on the type of film you use. How you intend to use the pictures will determine which format you need.

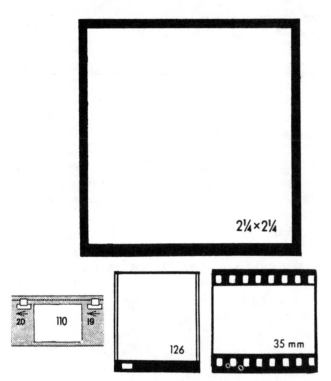

Fig. 9-2. Cameras can be classified by format—the size and shape of the image they make on the film. This shows the relative size of the negatives used in the more popular cameras today.

The 110 and 126 cartridge-loading cameras can produce snapshots or color slides with very little technical knowledge on the photographer's part. Low cost and ease of operation make these cameras good tools for teaching photographic composition to youngsters. However, the quality of prints from these cameras, especially the 110, is often unacceptable for most serious uses.

The 35mm camera is the nearest thing available to an all-purpose camera. It's the most popular choice for color slides. Black-and-white or color negatives from this format can usually be enlarged to 8 × 10 inches.

To make pictures for publications or exhibits, use a larger format camera, such as a 120 format which makes a negative 2¼ × 2¼ inches. This camera is not much larger than a 35mm, yet produces a negative nearly twice as large. Large negatives make sharper enlargements.

Recent improvements in instant picture cameras have increased their popularity. However, many experienced newspaper and publication editors won't use instant photographs because the lenses in these cameras are often too inferior to produce sharp pictures. The layers of color emulsion in instant color prints keep the edges of objects in the picture from looking sharp and crisp. If your newspaper prints a black-and-white picture from an instant color print, don't be surprised if the faces appear much darker than normal.

VIEWING SYSTEMS

Cameras are also classified by the viewing system, usually (1) viewfinder, (2) twin lens reflex, or (3) single lens reflex (see Fig. 9-3).

Viewfinder

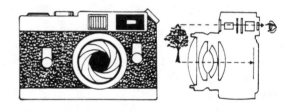

Twin Lens Reflex

Single Lens Reflex

Fig. 9-3. Cameras are classified by their viewing system, usually with viewfinder, twin lens reflex, or single lens reflex.

Viewfinder

The viewfinder camera has a window built into the top of the camera to allow you to compose your picture. It often contains an illuminated frame showing the margins of the picture. More expensive viewfinders contain focusing aids.

The viewfinder is excellent for general photography of scenery and people. It isn't well suited to extreme close-ups, due to the problem of parallax.

Parallax. When you don't look through the same lens the camera uses to take the picture, you see an image that is slightly different from what the camera records. This difference is called *parallax* (see Fig. 9-4). It's usually only a problem when you get closer than 5 feet from your subject. The closer the camera moves toward the subject, the worse the problem becomes. A single lens reflex camera completely eliminates this problem because you view and photograph through the same lens.

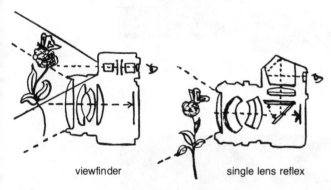

viewfinder   single lens reflex

Fig. 9-4. The photographer, who is taking a close-up with the viewfinder camera, thinks he/she will shoot the butterfly; however, the camera is taking a picture of the leaves below the insect. The single lens reflex camera allows the photographer to look through the same lens the camera uses. The photographer can see exactly what the photograph will contain, even on extreme close-ups.

Twin Lens Reflex

The twin lens reflex camera has two lenses, one directly above the other. The word "reflex" indicates that you view the image from a mirror inside the camera. The top lens is used for viewing and focusing, while the lower lens takes the picture. This camera is excellent for general print photography, but suffers from parallax in close-up situations.

Single Lens Reflex

If you plan to take many close-ups, such as of insects, stitchery, or graphics—or if you plan to use a wide-angle or telephoto lens—the single lens reflex is the most convenient viewing system. The photographer views the image directly through the lens by

means of a hinged mirror behind the lens. This completely eliminates the parallax problem. This system also makes it easier to focus and frame pictures when you are using a telephoto lens.

Single lens reflex cameras are available in most film sizes. Because they are more complex than other viewing systems, they are considerably more expensive.

CAMERA FEATURES

Once you have selected the format and viewing system, you'll find a variety of features on the cameras in the category you have selected. Each camera has a slightly different technique for focusing, adjusting the exposure, and loading the film. With practice you can usually adapt to any system.

There are, however, a few features you'll want to consider carefully. These include interchangeable lenses, built-in exposure meter, and automatic exposure control.

Interchangeable Lenses

This feature allows you to replace the normal lens with a specialty lens, such as a wide-angle or a telephoto. Nearly all cameras selling for $200 or more have this capability. It's a valuable feature which allows you to expand your photographic system as your needs and finances dictate.

Lens Speed

There can also be some variation in the normal lens you buy with your camera. For example, one option may be a 50mm, f/2, while the other choice may be a 50mm, f/1.4. The f-number in the lens description tells you how wide the lens will open on its largest setting. In this example the f/1.4 will let in more light (see section on f-stops) and is referred to as a faster lens.

Some people believe that since the faster lens costs more, it's a better lens. This is generally not true. In fact, the f/2 lens will usually produce a sharper image in all but the poorest lighting conditions. Unless you are interested in doing much low-light photography, you can save money buying the slower lens (in this example, the f/2).

Built-in Exposure Meter

Built-in meters have made handheld meters a thing of the past for most amateurs. Meters on most single lens reflex cameras read directly through the camera's lens.

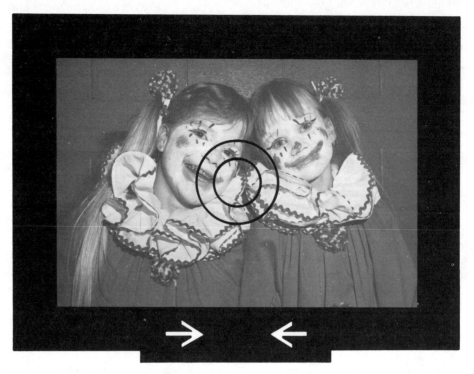

Fig. 9-5. Some very simple systems have dots or arrows that disappear when the exposure is correct.

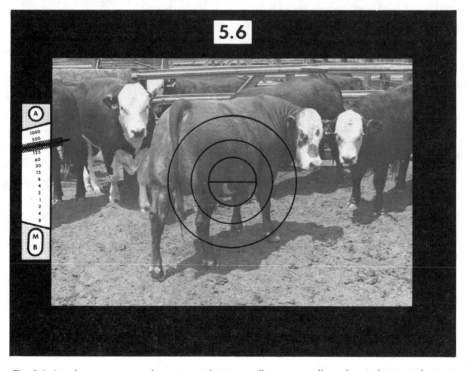

Fig. 9-6. In other systems, you have to match two needles or a needle and a circle to get the correct exposure. This model shows both the shutter speed and the f-stop.

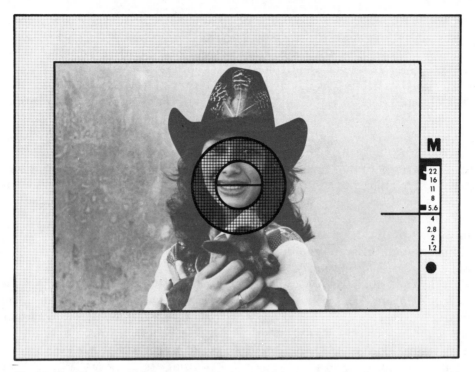

Fig. 9-7. The disadvantage of some automatic systems is that on a manual setting, you must reposition the camera by taking it down from your eye, and then you must set the lens opening on the number indicated by the needle in the viewing system.

To set the correct exposure, sight through the viewfinder and adjust the shutter speed or lens opening until the needle lines up on a certain mark. (See the examples in Figs. 9-5, 9-6, and 9-7.)

Some meters measure only a small portion of the subject being photographed, while others measure the entire picture area. Either system works well when you master its use. Through-the-lens meters are especially useful when you're taking close-up pictures and when you're using telephoto lenses.

Automatic Exposure Control

Automatic cameras are heavily advertised today. These auto systems carry convenience, complexity, and cost one step further by doing all exposure ad-

justments for you. The small gain in convenience over a built-in meter is hardly worth the additional cost. Automatic cameras can also produce poor photographs if you don't understand their limitations. (See section on exposure.)

If you decide to buy a fully automatic camera, be sure it also provides manual controls so you can still use the camera when the meter malfunctions or when the batteries are dead. Many automatic cameras use an electronic rather than a mechanical shutter, so they depend entirely on batteries.

You might also want to find out how much the replacement batteries cost. Since they sometimes run over $10, you might reconsider the simple match needle system. If you still want an automatic camera, avoid using it in extreme cold and always carry a spare battery.

◄ Selecting the Film ►

The film you select will depend on the kind of pictures you want to take. Before you buy your film, decide whether you want prints or slides, color or

black and white. You should also be aware of the lighting conditions—indoors, flash, or outdoors.

If you need black-and-white prints, use black-

and-white negative film, such as Plus-X or Tri-X. If you want color photographs, you must use color print film which always carries the word "color" in the name, such as Kodacolor II. If you want color slides, look for the word "chrome," such as in Kodachrome or Ektachrome.

SENSITIVITY TO LIGHT

Once you decide on the type of film you need, you must decide how sensitive the film should be to light. If you are taking pictures indoors without a flash or outdoors in the evening, you won't have much light. Conversely, a sunny day by the lake will be very bright. To accommodate such different conditions, film is produced in a range of light sensitivity that is rated by an ASA (American Standards Association) number, soon to be called an ISO (International Standards Organization) number.

For example, Tri-X has a rating of ASA 400 and is more sensitive than Plus-X, which is ASA 125. This means that it will take less light to record an image on Tri-X than it will on Plus-X. Tri-X (ASA 400) is a good choice for shooting low-light scenes in black and white.

On the other hand, films with higher ASA ratings also take on a grainy look, even though they may be in perfect focus. If you're taking pictures in bright sunlight or with an electronic flash, you won't need a highly sensitive film, so try to use a film with a lower ASA which will give you a finer grained picture.

The sensitivity of color film is also rated in ASA. However, if you use color film, you must take into account the type of light which will illuminate your picture. Most color films are balanced for daylight, which includes light from an electronic flash. You will get greenish pictures by using daylight film with common fluorescent light and a reddish-yellow cast by using daylight film with incandescent light. This off-colored cast can be reduced by using a corrective filter or an electronic flash.

READ THE INSTRUCTIONS

Film instructions are too often dropped into the nearest trash can without so much as a glance. But if you'll take a few minutes to read them, you'll probably take better pictures.

The instructions will tell you the film's ASA rating. By setting the camera's ASA dial, you're telling the camera how sensitive the film is to light (see Fig. 9-8). Once you set the ASA on your camera, don't change it until you load it with another roll of film.

The row of sketches on the instruction sheet tells you approximate exposure settings for different conditions. If your camera doesn't have a built-in

Fig. 9-8. Set your camera's ASA dial to match your film speed, and don't change it until you change film. Whenever possible, indicate somewhere on your camera the type of film your camera holds.

meter, cut out the diagram and tape it across the back of your camera (see Fig. 9-9). This will help you expose correctly and also will remind you what kind of film you are using.

The instruction sheet also explains how to take flash shots and difficult or unusual scenes, such as a stage show. If you use the film under unusual light conditions, it will tell you which filter to use.

Finally, the instructions urge you to keep film out of hot places, especially the glove box or window of your car. Film is sensitive to heat, as well as light, and a few hours in the car blast-furnace can ruin your pictures.

Because film is sensitive to heat, don't buy out-of-date film. While the film may still be good after the expiration date, its quality may decline. Your time is more valuable than the few cents you may save on expired film.

For short-term storage, refrigerate unopened film in moisture-proof plastic bags. Keep it in the door of

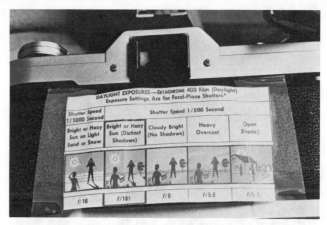

Fig. 9-9. If your camera doesn't have a built-in light meter, tape the exposure guide from the instruction sheet on the back of your camera.

the refrigerator away from spills. Before using the film, let it set at room temperature for at least three hours before opening the box.

◄ Taking Pictures ►

HOW TO HOLD THE CAMERA

The quickest way to assure sharp, clear pictures is to hold your camera correctly. Practice with an unloaded camera and hold it as if you're taking a picture. If you feel unsteady, you're probably holding the camera wrong.

Often people hold the camera along the edges and stick their elbows out (Fig. 9-10). They put their feet

WRONG

Fig. 9-10. This is a common but poor way to hold a camera. Holding the camera on the sides with your elbows out increases the chances that your picture will be fuzzy because the camera is not being held in a steady position.

together, lock their knees, and hold their breath before jerking the camera lever. This moves the camera every time, causing a blurred picture or cutting off heads. No wonder leading processors say that 90 percent of the bad pictures they process are due to one problem alone—camera movement.

Steady your camera by becoming a human tripod. Use your left hand as the tripod base, hold the camera as if it had been dropped into your hand. Wrap your thumb and forefinger around the lens, with the base of the camera resting on the palm of your left hand (see Fig. 9-11).

To complete the tripod, move the camera up to your eye, and let it rest on your cheekbone and nose. Operate the shutter release and the film advance with your right hand. This stance automatically pulls both elbows close to your body away from bumps. In this position it's only natural to spread your feet apart and relax your legs.

If you want to take a vertical picture, turn the camera in your hand with the film advance lever down, not up (see Fig. 9-12). This will keep your elbows at your sides where they belong. Holding the camera in this manner allows you to support most of its weight on the palm of your left hand. This leaves your right hand free to push the shutter release button and operate the film advance.

RIGHT

Fig. 9-11. Always support the camera in the palm of your left hand, with the thumb and index or middle finger on the focusing grip of the lens. This gives the camera firmer support and keeps your elbows against your sides.

RIGHT

Fig. 9-12. For vertical shots, turn the camera, with the camera lens resting in the palm of your left hand and your right elbow still against your side.

For extra support you can lean against a tree or a building. If you use a car for support, turn the engine off. You don't need the extra vibrations.

Frame your picture carefully in the viewfinder and you are ready to shoot. Relax. Keep both eyes open when you look through the viewfinder.

Just before you snap the picture, take a breath and then let it out. Yes, let it out, don't hold it in. This is much more relaxing than trying to tighten your chest enough to keep the air in.

As you release the shutter, keep a firm grip on the camera body. Move only your index finger in a slow, squeezing motion. Have someone stand in front of your unloaded camera as you practice taking pictures to tell you if you are jerking the camera.

If camera movement blurs your pictures, try resting your chin on top of your left-hand palm. This might feel a little ridiculous at first, but you'll soon see how much steadier it makes your hands.

ADJUSTING THE CAMERA

Before you can take a picture with an adjustable

camera, you have to adjust the size of the lens opening (the f-stop) and the amount of time the lens will be open (the shutter speed). These two settings regulate how much light hits the film. Most adjustable cameras have a built-in exposure meter with an indicator that you can see in the viewfinder. It tells you when the combination of these two settings is correct.

Usually photographers start by selecting a shutter speed. With a normal focal length lens, most photographers will start with a 1/125 shutter speed for photographing inanimate objects.

After setting the shutter speed, set the lens opening. The indicator in the viewfinder will tell you a number to set on your lens, or it will tell you when the setting is correct by matching two needles.

Shutter Speeds and f-stops

If you look carefully at the shutter speed dial (see Fig. 9-13), you'll notice that each number is about twice as big as the number on one side and half as big as the number on the other side. Since these are fractions of a second with the numerator (1/) left off, it's easy to realize that a 250 setting lets light pass through the lens twice as long as a 500 setting and one-half as long as a 125 setting.

The same general principles apply to the numbers on the lens barrel (see Fig. 9-14). Think of these numbers as fractions with the 1/ left off. This will help you remember that f/16 is a much smaller lens opening than f/1.4 (see Figs. 9-15 and 9-16). The

Fig. 9-13. Each number on the shutter speed dial is twice as large as the number on one side and one-half as large as the number on the other side. These are fractions with the numerator (1/) left off. A 250 setting lets light pass through the lens twice as long as a 500 setting and one-half as long as a 125 setting.

Fig. 9-14. The lens-opening numbers follow the same principles as the shutter speed dial. The larger numbers let in less light, and each opening is one-half as large as the one before it.

Fig. 9-15. Think of the f-number as a fraction with the 1/ left off. This will help you remember that f/16 is a much smaller lens opening than f/1.14.

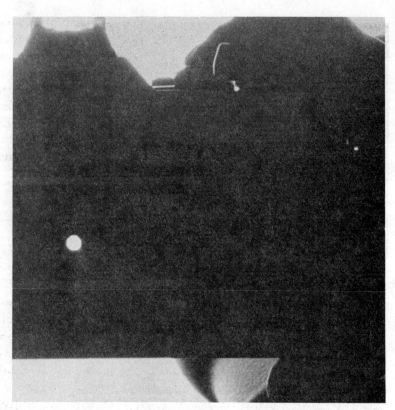

Fig. 9-16. The problem with thinking of f-numbers as fractions is in the math—f/11 doesn't seem to be twice the size of f/16—but it is.

problem with thinking of them as fractions is in the math—f/11 doesn't seem to be twice the size of f/16, but it is. This is because the lens opening numbers refer to the diameter of the circular lens opening divided into the distance from the lens to film plane.

Rather than figuring the math, just remember that the lens opening ring is set up exactly like the shutter speed dial. An f/11 lens opening lets in twice as much light as an f/16 and one-half as much light as an f/8.

This makes changes in the camera setting very simple. Say you are taking pictures at 1/60 and f/8. Suddenly you need to change to a 1/250 to stop action. You have cut the length of time the light can pass through the lens in half two times or two stops. To get the same total exposure, you need to double the size of the lens opening two times, so you move the lens opening from an f/8 to an f/4 (see Fig. 9-17).

Of course, you don't always have to use a fast shutter speed to show action. A slow shutter speed, causing a blur on part of your picture, is often much more effective (see Fig. 9-18).

EQUIVALENT LIGHT SETTINGS

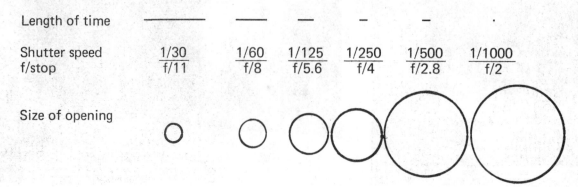

Length of time	———	——	—	–	–	·
Shutter speed	1/30	1/60	1/125	1/250	1/500	1/1000
f/stop	f/11	f/8	f/5.6	f/4	f/2.8	f/2
Size of opening						

Fig. 9-17. Study this ratio between the length of time the light can pass through the lens to the size of the lens opening. One step of one setting equals one step in the other setting.

Fig. 9-18. You don't always need a shutter speed fast enough to stop the action. A blur, as in this picture, can be quite effective.

Depth of Field

But what about lens openings? Are there other reasons to change them besides when you need to compensate for changing shutter speeds?

Let's think of an example that works in everyday life. If you see a child squinting to see the chalkboard, you would know the child is having trouble focusing and probably needs glasses. By squinting, the child is cutting down on the amount of light entering the eye and is making more things appear to be in focus.

To put this in photographic terms, if you can make the camera "squint" by using a smaller lens opening, the area which is in sharp focus gets larger. Photographers say, "The depth of field increases."

If you are taking a picture of a man vaccinating an animal, you may want both the needle and the man's face to be in focus. That's no problem if both are about the same distance from your camera. More likely though, the needle is closer to the camera than his face is. Then you need sharpness not just at one distance but through a range of distance, from 5 to 7 feet or more. You gain this depth of field by using a small lens opening, or f-stop.

Most cameras have a depth-of-field scale engraved on the lens' barrel. This scale is a set of lens-opening numbers which appear on each side of the focusing mark. By looking at the f-stop number you have selected on both sides of the scale, you can see the range of distance which will be in focus.

For example, say the man's face is 7 feet away. By using a lens opening of f/16, you'll be able to focus on everything from 5 feet to about 12 feet, including the needle which may be at 5.5 feet (see Fig. 9-19).

Your next picture may be of just the needle without emphasizing the man's face. To take this picture you use an f/4 or f/2.8 aperture. At this setting the depth of field is much smaller. If you focus on the needle at 5.5 feet, the face at 7 feet will be outside this depth of field and out of focus (see Fig. 9-20).

But as you look through your camera, you may not be able to see this depth of field, even after you've set the smallest lens opening. This is because your camera lens remains all the way open until the split second you take the picture.

Fig. 9-19. On an f/16 setting, the depth-of-field guide tells you that everything between the two 16's (5 to 11 feet) will be in focus, so both the needle and the man's face are focused.

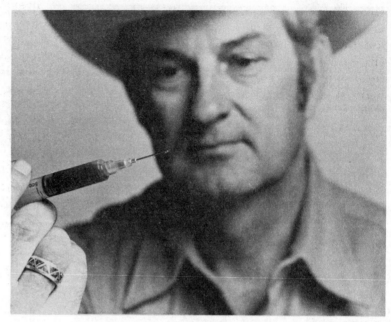

Fig. 9-20. With an f/4 lens opening, only the area very close to your point of focus will appear in focus. This means the man's face will be out of focus if you focus on the tip of the needle.

Cameras with a depth-of-field preview button allow you to close the lens down to the selected setting and permit you to check the focus on the man's face and the needle. Closing down (or stopping) the lens can make the viewing screen quite dark, so you do it only while you check the depth of field. Then, before taking the picture, open the lens to allow careful viewing of the scene.

EXPOSURE

With an automatic camera it's easy to take exposure determination for granted and leave all the control up to the camera. Automation works fine for average subjects, but it can lead to poor pictures if the subject is extremely dark or light, or if the lighting is contrasty.

A light meter, whether built into the camera or handheld, is an instrument that interprets everything it's pointed at as a middle shade of gray. If you're photographing a green field with the sun overhead, following the meter reading will give the correct exposure because that scene approximately matches the middle shade of gray reflectance. But if you try to photograph a snow-covered landscape, following the meter will produce an underexposed picture. The meter will still interpret the scene as the middle shade of gray, even though the snow scene reflects about two stops more light.

With light or dark subjects or contrasty lighting, try to base your meter reading on the area in which you want to record significant detail, rather than on the entire scene. You may have to make a close-up reading or a substitute reading off the palm of your hand to get the correct result.

To make a substitute reading, just hold your hand with the palm facing your camera in approximately the same lighting conditions as your subject. Set your exposure for the light reflected from your palm, but then open your lens one f-stop more than that indicated by the meter because your palm is approximately one stop lighter than the middle shade of grey. For example, if your off-the-palm reading suggests using f/8, you should open the lens to f/5.6 before taking your picture. This technique takes longer to explain than it does to use in practice, but it's a very helpful technique for difficult exposure conditions.

Bracketing your exposure is another technique that professional photographers use when faced with unusual lighting conditions. Bracketing means shooting a series of pictures at different exposures rather than making just a single exposure of your subject. Begin by making your first exposure at the setting recommended by your meter or best guess. Take a second picture, increasing the exposure by one stop, and a third, decreasing your exposure by one stop less than your first. This process gives you three chances of getting a correctly exposed picture instead of just one. For critical work with color transparency film, bracket in one-half f-stop increments rather than whole stops. This process requires you to use more film, but it's the best guarantee of getting a good photograph.

AUTOMATIC EXPOSURE CONTROL

There are two automatic systems—shutter priority and aperture priority. Shutter priority means the shutter speed is most important, so that is the control you set. The meter adjusts the lens opening for proper exposure at the shutter speed you've selected (see Fig. 9-21).

The advantage of this system is that you can select the shutter speed you need to freeze or blur the action. The disadvantage is that the camera may select an f-stop that gives you either too much or too little depth of field for the composition you desire.

Fig. 9-22. Aperture priority systems have the "Auto" on the shutter speed dial. You set the f-stop, and the camera selects the shutter speed.

Fig. 9-21. Shutter priority systems have the "A" symbol on the f-stop ring. You set the shutter speed, and the camera selects the f-stop needed to produce an average exposure.

The aperture priority system works just the opposite. You select the f-stop or lens opening for the desired depth of field (see depth of field), and the camera selects the shutter speed to give you the right exposure (see Fig. 9-22).

The disadvantage is that the camera may drop your shutter speed below the speed at which you can successfully handhold the camera. It's difficult to hold the camera still below 1/60 second, with a short or normal lens.

Either the shutter priority system or the aperture

priority system will give you properly exposed photographs. But you should be aware of the settings the camera has selected. If you aren't, your automatic camera could give you properly exposed "blurs," or pictures with a distracting depth of field.

Some expensive automatic cameras select both the shutter speed and the lens opening. But even these cameras can't read your mind and don't know when you want to use an unusual setting to create a certain effect.

Estimating Exposure

If you don't have a built-in light meter or a separate handheld light meter, use settings suggested on the film's instruction sheet. Tape them on your camera as a guide.

You can also try the "Sunshine Rule of 16." This rule of thumb works with any kind of film for outdoor shots. The rule is: "When shooting under bright sunshine, simply set your f-stop or lens opening on f/16. Then adjust your shutter speed to the speed closest to the ASA of your film."

This means that with ASA 400 film, you would set your lens opening on f/16 and use a 500 (1/500 of a second, that is) shutter speed. With Ektachrome 64 film, again use an f/16 lens opening with a shutter speed of 60.

A photograph can be an exciting, dynamic visual statement or simply a dull record of what was happening when the shutter was released. The difference is usually not in the quality of the camera used or in luck, but rather within the creativity and motivation of the photographer. In order to make photographs that communicate effectively, the photographer must know how the picture is to be used and what is to be communicated. An attention-getter photograph for an exhibit, for example, may have very different requirements than a photograph accompanying a news story, yet both pictures could probably be made at the same location.

The examples in this chapter are the same types of events and situations you often must photograph. Study these examples when you need new ideas for standard shots.

PICTURE USES

News photos. To illustrate a news article, keep the picture simple and get as close as possible to your subject. Avoid handshakes and shots of people holding trophies (see Figs. 9-23, 9-24, and 9-25). If the photo is to report on someone receiving an award, take a picture of the recipient doing whatever he/she did to earn the award rather than a picture of the award ceremony (see Fig. 9-26). Arrange news

Fig. 9-24. The photographer taking this check-presentation picture improved the visual impact by placing the student in front of the donor, bringing the points of interest closer, and throwing the donor slightly out of focus, thus emphasizing the student receiving the award.

Fig. 9-23. The typical grip-and-grin picture may be a nice souvenir for the person receiving the award, but it isn't a news photo. The problem is compounded even further when a third person, who is neither giving nor receiving the award, is added. Avoid these conditions whenever possible.

Fig. 9-25. One solution is to move in close to catch the reflection of the winner's face in the plaque. You can read the plaque and see who won it at the same time.

Fig. 9-26. The best way to photograph an award winner is in the activity he/she conducted to deserve the award. For example, photograph a researcher in the lab, or the farmer-of-the-year in the field.

photos to include as few people as possible and select a camera angle that shows faces.

Check with managing editors in advance about deadlines and the type of film they prefer. Some papers will use Polaroid prints, and others may agree to develop your film for you.

Displays and exhibits. Photographs are a good device for calling attention to exhibits and displays. The story should be evident at a glance, so select pictures that are simple and have dramatic compositions. Unusual camera angles, dramatic action, and extreme close-ups are sure eye-catchers (see Figs. 9-27 and 9-28). The larger the pictures, the better. Use nothing smaller than 8 × 10 inches, with 11 × 14 or 16 × 20 inches ideal. Both color and black-and-white prints can be effective, but black-and-white prints will usually show up better from a distance. Prints should have a mat finish, so they are not shiny and reflective.

Reports and records. Photographs often illustrate various points in a report better than words. If the report isn't to be printed, use dry mounting tissue or other photo mounting adhesive to attach the print to the page. Avoid using photo-mount corners; they usually fall off with handling. Also avoid water-base glue which will cause the print to buckle.

Fig. 9-27. Imagine where the photographer had to stand to get this shot. Yet, the effective use of foreground and corn flying towards the viewer makes this an interesting, new approach to the standard corn harvest picture. This is an ideal photograph for a display.

Fig. 9-28. The extreme close-up, coupled with the out-of-focus viewer, conveys a complete story. The selective focus keeps the face from distracting from the grasshopper. The fingers holding the grasshopper help draw the viewer's eye to the picture.

COMPOSITION

When you pick up a camera and click the shutter, you will record an image on the film. Whether that image is good or bad depends on how you organize the subject through the viewfinder. We call this organization composition. There are no hard and fast rules as to what is good or bad composition. But these guidelines will usually improve your pictures.

1. *Center of interest.* When you look at a photograph, your eyes should immediately recognize one thing as the most important element of the picture. That element is called the center of interest (see Fig. 9-29). Your eye will

WEAK

STRONG

Fig. 9-29. Faces almost always make strong centers of interest. You can strengthen the impact of a photograph in some cases by including enough surrounding area to allow for an off-center placement of your subject, especially if the surroundings are uncluttered, but yet enhance the story line of the photograph.

Fig. 9-30. The eye searches for the bright spot in a predominately dark picture. Here the eye moves toward the *Seedway* magazine and the man's nose. This mailbox shot was taken with the camera in the back of the box with a 20mm lens. To produce this shot, the photographer took a meter reading on the shadows of the man's face, prefocused by the distance from the back of the box to the man's face, set the self-timer, placed the camera in the box, and waited—about 8 seconds.

always gravitate toward the lightest area in the photograph (see Fig. 9-30). Try to compose your picture with the center of interest in a well-lighted area. Keep the picture simple; leave out all unnecessary details. If you want to show three things, take three pictures.

2. *Rule of thirds.* The best place for the center of interest is usually not exactly in the center of your picture. Try dividing the viewfinder into thirds, using vertical and horizontal lines as shown in Fig. 9-31. Compose your pictures so the center of interest falls where two of these lines cross, and it will usually be more interesting to look at than if it was centered (see Figs. 9-31 and 9-32). Try to keep the horizon line in the upper or lower third of your picture and not through the center. Try to arrange your subject matter for balance but not necessarily symmetry.

3. *Depth.* You can create a feeling of depth by composing a picture with an object closer to the camera than the center of interest. For instance, framing through tree branches or placing another object in the foreground adds depth to a landscape picture (see Figs. 9-33 and 9-34).

Fig. 9-31. Mentally divide your viewfinder into thirds, both horizontally and vertically. Compose your picture so the center of interest lies where these lines intersect, rather than dead center. This will usually give your pictures a more dynamic and exciting look because the human eye quickly becomes bored with symmetry.

Fig. 9-32. Foreground helps put the subject into context. By crouching in the wheat field, the photographer was able to get a low, strong angle of both the wheat and the farmer.

Fig. 9-33. Foreground which surrounds the subject can have a framing effect.

WRONG

RIGHT

Fig. 9-34. Notice how much more information and impact a picture contains when the foreground and framing are used wisely.

4. *Action.* Keep people busy, and they will appear more relaxed and natural. Avoid having your subject staring directly into the camera unless it's a formal portrait. A few interesting props give the subject something to work with and help create a natural feeling (see Figs. 9-35 and 9-36).

5. *Camera angle.* An unusual camera angle or viewpoint can add a great deal of interest to an ordinary subject (see Figs. 9-37, 9-38, and 9-39). A low camera angle suggests strength and drama. A high camera angle will make your subject look more submissive and gentle.

Fig. 9-35. Keep people busy, but be sure you can see most of their faces.

Fig. 9-37. This low angle increases the strength and impact of an ordinary scene.

Fig. 9-36. To capture the subject's facial expression in a dark area outdoors, use a flash with a diffuser. Notice how the flash adds snap to the picture without making the lighting look artificial.

Fig. 9-38. Avoid taking over-the-shoulder pictures from the middle of the audience. This angle produces dull, uninteresting pictures.

Fig. 9-39. Instead, move close, using a wide-angle lens. This allows you to include both the speaker and the audience. As well as being much more interesting, this angle makes the viewer feel like a part of the activity.

6. *Background.* Unless it definitely tells part of the story, keep the background plain (see Figs. 9-40 and 9-41). Avoid extremely light or dark backgrounds.

7. *Timing.* While timing is one of the most important techniques used by professionals, it is often overlooked by amateur photographers. Try to get your picture at the most dramatic moment, the one that symbolizes the event you're photographing (see Figs. 9-42 and

Fig. 9-40. Cut clutter from your picture by using the sky as a background. This lower angle also accents the icy pond and the motion of the skaters.

Fig. 9-41. Backgrounds can also add to the story line of pictures. The presence of cattle in the background of this photograph reinforces the discussion of the two men on the economics of livestock production.

9-43). This is just as important when taking a picture of someone working as it is when photographing a sporting event. Train your instincts to respond to the most significant moment when you are taking pictures of someone. It often requires shooting several pictures rather than just one or two.

8. *Leading lines.* Look for lines which will draw the viewer's eye directly towards your center of interest. Note how both pictures in Fig. 9-44 look very dull, especially when compared to the impact they contain when the photographer adds creative leading lines (see Fig. 9-45).

9. *Selective focus.* Selective focus can make a big difference in the message your picture conveys. Fig. 9-46 uses a small lens opening to increase the depth of field. Since the house is in focus, it is a major picture element. The message seems to be "poor rural housing conditions." A 3.5 lens opening in Fig. 9-47 knocks the house entirely out of focus to emphasize the pump. The message here is "old, artistic-looking pump." Use of a soft focus for the background removes distracting elements, such as the dog, thus strengthening the composition. (See depth of field.)

10. *Formal portraits.* These pictures are often produced in a studio. If you have flash equipment, a simple setup, used in Fig. 9-48, involves a diffused electronic flash held above and left of the subject. The 100mm lens is ideal for this work. Also consider taking portraits outdoors on an overcast day (see Fig. 9-49). The clouds diffuse the light, creating

Fig. 9-42. Try to shoot pictures at the most dramatic moment when the action is at its peak and the subject is best symbolized. Compare this picture of hay baling with Fig. 9-43 taken a few seconds later.

Fig. 9-43. Here the photographer missed the peak of the action. The hay bale is already down, the closest person's face is in the shade, and one person is standing in front of another.

shadowless illumination. Don't overlook the possibility of taking the subject in a natural setting. Fig. 9-50 was shot with only the room and lamp light. It conveys much more of the artist's personality and occupation than Fig. 9-51 does.

◄ Lighting the Scene ►

The world is full of light sources for your photographs. You can use a camp fire, a light bulb, the moon, a candle, reflected light from a white wall—the list is almost endless. Of course, the sun is about the most common source, and its quality changes almost hourly—from the soft luminescence of pre-dawn to the hard brilliance of noon and the orange glow of sunset. The same scene shot at various times during the day can take on a completely different look (see Fig. 9-52).

High speed films make it possible to take indoor shots without using a flash. For the most natural look, try to arrange for side lighting, possible from a near-by window or open door.

If you use daylight color film under artificial lights, you will get poor color rendition. This can sometimes be improved with the mixture of sunlight or light from a flash.

Using a flash allows you to select a fine grain, slow speed film. You will also gain depth of field and easily stop action with a flash.

CAMERA SETTING FOR FLASH

When using an electronic flash, you no longer need fast shutter speeds to stop action. In fact, the duration of the flash is shorter than the fastest shutter speed, which means the flash itself will stop almost any action, regardless of the shutter speed set on the camera.

It's very important to set the shutter speed so it will be synchronized with the flash. If you don't, your photograph will appear to be cut in half, with one part exposed properly and one part black (see Fig. 9-53).

Most cameras indicate which shutter speed you

Fig. 9-44. Often there are lines which can be included that will draw the viewer's eye directly toward the center of interest. Both of these pictures are very dull.

Fig. 9-45. The impact of these same two scenes is increased when the photographer adds creative leading lines.

Fig. 9-46. A small lens opening increases the depth of field, thus bringing the house into focus and making it a major picture element.

Fig. 9-48. If you have flash equipment, you can make a simple studio setup, with a diffused electronic flash held above and left of the subject.

Fig. 9-47. A 3.5 lens opening knocks the house out of focus and emphasizes the pump.

Fig. 9-49. Attractive portraits can be taken outdoors on an overcast day.

Fig. 9-50. This low-light shot was taken with only the room and lamp light.

Fig. 9-52a. You should experiment with light coming from all directions, especially back and side lighting. The late afternoon back lighting on these soybeans helps show the pubescence of the pods.

Fig. 9-51. This better-lit portrait doesn't reveal the artist's personality and occupation as well as did Fig. 9-50.

Fig. 9-52b. Shooting into the sun can create separation and depth, thus bringing about a three-dimensional illusion. You must be sure that no direct sunlight falls on the lens. A lens shade will shield the lens and enhance the contrast range. Take a meter reading from the shaded side of the silo.

should use by making the number (usually 1/60 or 1/125) a different color, or it will be marked with a symbol (see Fig. 9-54). Older cameras have two outlets for the flash cord—one for bulb and one for electronic units. The connection for electronic flash is usually marked with an "X."

Many cameras today have a hot shoe attachment in the top so you don't need to use a flash cord. Some cameras even automatically synchronize the camera when the flash is attached.

Since the shutter speed remains constant, you must change the lens opening to adjust the exposure. Subjects near the flash receive plenty of light, so you can use a small lens opening. Subjects further away from the flash receive less light. Therefore, the lens must be opened wider to take in enough light for a proper exposure.

To determine the correct lens opening, use the

Fig. 9-53. If you don't use the correct shutter speed, your flash picture will look like it has been cut in half. This picture was shot at 1/125 when the camera indicated 1/60 should be used with flash.

Fig. 9-54. Most cameras indicate the proper shutter speed to use with a flash by the correct speed on the dial.

Fig. 9-55. Use the scale on the flash to determine the best f-stop. Set the ASA on the dial and use the f-stop across from the distance. For this particular unit, use f/22 when your subject is 10 feet away, and f/11 when he/she is 20 feet away.

scale on the back of your flash unit. For manual flash units, simply set the ASA on the flash, and use the suggested lens opening (see Fig. 9-55).

Many flash units have automatic exposure features which eliminate the need for continuous exposure adjustments. When the flash unit is set on automatic, its light sensitive eye will measure the light reflected from the subject and control the duration of the flash (see Fig. 9-56).

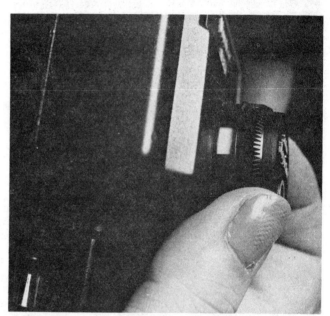

Fig. 9-56. Automatic flash units have a sensor eye which can be color-coded with a scale on the flash to adjust the light output automatically, based on the amount of reflected light.

A computerized automatic flash is great for taking fast action pictures. The key is to set the flash and camera for the maximum distance you expect during the shooting session. From that point on, you need only to focus and shoot. The unit's electric eye will adjust the exposure according to the light reflected from the subject.

Another advantage of an automatic flash is that it allows you to choose from a range of f-stops to control depth of field.

USING FLASH OUTDOORS

In bright sunlight a flash can reduce deep shadows and lighten the subject.

To figure the correct exposure, remember that you must use the shutter speed which synchronizes with the flash. Check the light meter reading for the lens opening at that shutter speed. Then adjust your distance from the subject until the lens opening agrees with the opening suggested by the scale on your flash.

Fig. 9-57. The hot shoe attachment is the most convenient place for the flash, but the picture often loses the soft gray tones needed here on the nose and forehead.

Fig. 9-58. By using a side bracket and bouncing the light off a white card, you can return the gray tones to the face, making the picture more normal.

FLASH ACCESSORIES

The most common placement of the electronic flash is on the camera's hot shoe which can carry the electric impulse directly to the flash without the use of a cord. While this placement is very convenient, it produces a picture that looks flat and unnatural (see Fig. 9-57).

In nature, subjects are rarely lit from the front but almost always from the top or the side. To make your subject look natural, try to light it from a side angle. Side lighting will also be more comfortable for your subject and will eliminate red eyes on your subjects.

You can purchase a bracket to hold your flash to the side and slightly above the camera. This requires a connecting cord from flash to camera but helps provide better lighting effects.

You can also solve the problem of harsh, unnatural shadows by using a diffuser, or by bouncing the light off a reflective surface (see Fig. 9-58).

A plastic diffuser fits over the flash's reflector to soften the light. Diffusers are available with various grid patterns for use with normal, wide-angle, or telephoto lens.

You can make your own inexpensive diffuser by taping several layers of white tissue or matte acetate over the flash unit (see Fig. 9-59). Use this rule of thumb for a manual flash unit—open one-half f-stop per layer of tissue—to figure your exposure.

Another way to diffuse light is to bounce the flash off a nearby white or almost white wall or ceiling. While there are formulas to figure the exact lens

Fig. 9-59. Make you own inexpensive diffuser by taping several layers of white tissue over the flash. Open the lens one-half f-stop per layer of tissue.

opening when you "bounce" the flash, in general open your lens two f-stops to compensate for the decreased illumination.

Of course, the bouncing technique is much simpler if you have a computerized automatic flash unit.

◄ Camera Accessories ►

PROTECTING YOUR CAMERA

The first camera accessories you need are those which will protect your investment. These include protective filters, cleaning supplies, and protective bags.

Protective filters. Filters protect lenses from scratches, bumps, and dents. Since lens paper can often scratch the lens, buy a filter with each lens so you rarely have to clean the lens. Instead, you will clean your protective filter when needed.

Protective filters can be purchased under various names, including haze, ultraviolet (UV), or skylight.

These filters will also cut haze or ultraviolet light reflections.

Cleaning supplies. To get the most benefit from your camera system, you must always keep it clean. Carry a supply of lens tissue and fluid. Don't confuse this with eyeglass tissue which will scratch the lens (see Fig. 9-60). The silicone material contained in eyeglass tissue will also ruin the antireflection coating on your lens. Carry a soft brush to wipe away dust, and an ear syringe or a can of compressed air to blow dust from the internal parts of the camera.

Protective bags. Inexpensive gadget bags will protect

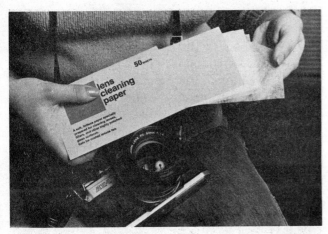

Fig. 9-60. The only paper you should ever use to clean your lens is the type designed for photographic lenses.

and organize your equipment. These spacious bags often have compartments for lenses, film, and accessories (see Fig. 9-61).

A photographer who travels extensively or who works in extremely dusty situations needs a good case, preferably foam-lined. While these cases are

heavier and don't carry much equipment for their size, they are a valuable protective devise for your camera equipment (see Figs. 9-62 and 9-63).

Many photographers who plan to always carry a gadget bag or carrying case don't buy the tight-fitting leather camera cases.

FILTERS

Filters can be important when you are using black-and-white film. Even though colors such as red and green look quite different to your eye, they may appear as the same shade of gray in a black-and-white print.

To make the black-and-white print more natural looking, you can use filters to darken some colors and lighten others. Colored filters lighten colors similar to their own and darken colors that are complementary (see Fig. 9-64).

In black-and-white photography, a yellow filter will make yellow and orange appear lighter, while blue will appear darker. As a rule, the most useful filter for black-and-white photography is a

Fig. 9-61. Inexpensive gadget bags will protect your equipment and help you keep it organized.

Fig. 9-62. This more expensive, foam-lined bag is better for the photographer who travels extensively or who works in dusty conditions.

Fig. 9-63. An army ammunition box makes an inexpensive, rugged, dustproof, waterproof box for your camera gear. Glue foam rubber to the interior to hold and protect your equipment. Painting the exterior white or silver to reflect sunlight will keep your equipment cooler.

yellowish-green filter which darkens a blue sky so that white clouds stand out.

A polarizing filter will help reduce unwanted glare and haze, darken blue skies to make the clouds stand out, reduce reflections in glass, and make the colors appear richer. This filter is valuable for both black-and-white and color photography.

When you are using color film, three filters will interest you. No. 80B makes it possible to use daylight color film with photolamps 3400° Kelvin. No. 80A gives correct color rendition when you are using daylight film with regular incandescent lights or

FILTERS for BLACK-and-WHITE Photography

SUBJECT COLOR	FILTER TO LIGHTEN THIS COLOR	FILTER TO DARKEN THIS COLOR
Blue	Blue	Red, yellow, orange
Green	Yellow, green, orange	Red, blue
Yellow	Yellow, green, orange, red	Blue
Orange	Yellow, orange, red	Green, blue
Red	Red	Blue
Purple	Blue	Green

Common filter names based on Kodak Wratten designations: light yellow, 3 (K1); medium yellow, 8 (K2); light green, 11 (X1); dark green, 13 (X2); dark yellow or orange, 15 (G); red, 25 (A); blue, 47 (C5).

Fig. 9-64. Filters to darken some colors and to lighten others make black-and-white prints more natural looking.

tungsten lights. When you are using tungsten-balanced film indoors with a flash or outdoors without a flash, use an 85B filter.

There are six common types of fluorescent light, and each requires different filter combinations for optimum results. For less critical work you can use an FLD filter and daylight film. If you add an FLD filter, open your lens one stop.

There are various special effect filters you may wish to try. A starburst filter will create four-pointed stars on any bright point of light. A neutral density filter allows you to use high speed film in bright light. Also try a repeating or multiple-image filter-type lens.

LENSES

Lenses are generally classified as normal, wide-angle, and telephoto (see Fig. 9-65). A normal lens

Fig. 9-65. The main difference between various lenses of the same brand is the focal length—the effective distance between the lens and the film when the lens is focused. For most 35mm cameras, a 50mm lens is standard. Any lens with a focal length longer than this is called a telephoto lens and will bring the subject closer, while any shorter than 50mm is considered a wide-angle lens and will take in more of the scene.

has a focal length approximately equal to the diagonal of the negative. A 50mm lens for a 35mm negative is considered normal, and for a 2¼- × 2¼-inch camera, 85mm is considered normal. A normal lens has an angle of view of about 50 degrees, which is similar to the angle that the human eye sees sharply.

Lenses with a focal length shorter than the negative diagonal are considered wide-angle, while the longer ones are usually called telephoto. The longer focal length lenses cover a smaller area and compress the apparent distance between the foreground and background. The shorter focal length lenses cover a wider area and tend to distort subjects near the camera.

As a general rule, the slowest shutter speed you can safely handhold equals the lens length. This means that for a 100mm lens, you should use a shutter no slower than 1/125 when handholding.

Zoom lenses (lenses with variable focal length) often appear to be the perfect compromise. But while these lenses may be fun to use, they are often too heavy and bulky to carry permanently on a camera.

Before you buy extra lenses, exploit your standard lens fully. Soon you'll know what type of pictures interest you most, and you can buy a high quality lens for this purpose.

Lens Selection

When selecting lenses, try to limit your choice to as few as possible. This will save money, extra weight, and indecision over which lens to use. Most experienced photographers settle on one or two favorite focal lengths for the majority of their work and reserve other focal lengths for special purposes. For photojournalism, a moderate wide-angle and a short telephoto lens are most useful. In a 35mm format, this would be a wide-angle of 35mm focal length and a telephoto of 100mm focal length.

The wide-angle lens is ideal for pictures of two or three people engaged in a common activity, and for pictures of scenery, where you need extra depth of field (see Fig. 9-66). The telephoto lens is useful for portraits and close-up shots because it produces a more pleasing perspective than the normal lens. The shallower depth of field inherent in the telephoto lens makes it possible to throw backgrounds out of

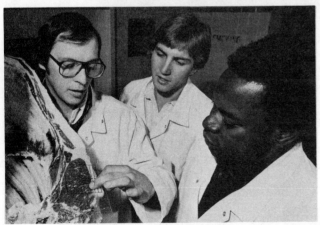

Fig. 9-66. A telephoto lens (135mm in this case) compresses the scene, accentuating the closeness of the houses to the farmland. A mid-range wide-angle (35mm) in the meat lab works well in tight quarters with little distortion, while a superwide-angle (16mm or fisheye) takes in a very wide view of this biochemical lab, yet tends to distort everything near the edge of the picture and near the lens.

focus, concentrating your viewers' attention on the main subject.

Once you have these lenses, you'll find that you use the normal lens so seldom that it hardly pays to own one. In fact, if you're purchasing a new camera, you would probably save money buying just the camera body with the moderate wide-angle and the telephoto and not buying the normal lens at all.

Use a lens shade on any lens that is not deeply recessed in the lens barrel. The shade prevents glare and protects the lens from bumps.

OTHER ACCESSORIES

Extra batteries. Batteries in any camera are important, but they become critical when you're using a fully automatic camera. Many of these cameras depend on the battery to operate not only the light meter but also the lens opening and the shutter. A dead battery means no pictures.

Lights. You need these for copy work and indoor shooting. Clamp-on lights with reflectors are inexpensive and flexible. Copy stands offer even more versatility. Select lamps of the color temperature Kelvin that matches the film you will be using. (See section on close-ups.)

Tripods. These steadying devices permit the use of slow shutter speeds and small lens openings for greater depth of field. Tripods are also sometimes used for copy jobs. Select a sturdy tripod that's easy to set up and adjust. It should have a center post for quickly changing camera height and an easy-to-operate, positive-locking pan-tilt head. Metal tripods are generally more rigid and durable than those with plastic parts—stability is the crucial feature here.

You will often need a cable release to keep from moving the camera when you release the shutter.

Autowinders and motor drives. These accessories are popular among sports and portrait photographers. They release the shutter, advance the film, and cock the camera.

Autowinders advance the film at the rate of one or two frames per second. Motor drives advance up to five frames per second and are heavier than autowinders.

Film holders and protectors. If you'll be using several rolls of film in one outing, film canisters on your camera strap will help you keep your hands and pockets free. They also protect against misplaced or damaged film (see Fig. 9-67).

A lead-shielded film bag is a good idea for photographers who frequently pass through airport X-ray checks.

Fig. 9-67. Not only is a film holder on your camera convenient, but it also decreases the chances that you'll leave the film in a hot car or glove box.

◄ Close-up Photography ►

Nearly everyone who uses a camera can think of occasions when it would be important to take close-up photographs. Close-ups add greatly to the impact of a slide show and can be a valuable teaching tool as well. Close-up photography isn't difficult but does require some additional equipment and practice.

CLOSE-UP DEVICES

There are two techniques for adapting a single lens reflex camera (SLR) for close-up work. One is to add a supplementary (close-up) lens to the camera's existing lens. The other is to increase the distance between the lens and the film plane by using extension tubes or bellows.

Supplementary Lenses

Supplementary lenses are the least expensive and most convenient solution. They attach directly to the front of the camera lens like a filter, converting its entire focusing range to shorter distances.

Supplementary lenses are really low-power magnifiers and are available in diopter strengths from +½ to +10. The higher the number, the closer you can focus. A +2 and a +4 supplementary lens are a very useful combination for a 35mm camera equipped with a 50mm normal lens. Supplementary lenses stronger than +5 are quite difficult to use and may reduce the sharpness of your picture. The areas you can photograph using various close-up lenses on a 50mm lens are shown as follows:

Supplementary (close-up lens)	Camera focusing scale set at infinity	Camera focusing scale set at 24 inches
+2	9 × 13½ inches	4½ × 6½ inches
+3	6 × 9 inches	3¼ × 5 inches
+4	4½ × 7 inches	2¾ × 4¼ inches

The advantages of supplementary lenses are:

- They are small, light, and easy to carry in the field.
- They require no exposure compensation.
- They can be used on cameras without interchangeable lenses.
- Two can be combined to get even closer to the subject.
- They don't interfere with the automatic diaphragm function of SLR lenses.

Extension Tubes and Bellows

Extension devices can be used only on cameras having interchangeable lenses. They are inserted between the lens and camera to increase the lens-to-film distance. The greater the extension, the closer the camera will focus.

Extension tubes are rigid metal rings that can be used separately or in combination to achieve varying amounts of extension. When used with a 50mm lens, tubes are best suited to photographing subjects one inch high or larger.

The advantages of extension tubes are:

- They are less expensive than bellows.
- They are fairly compact and rugged, making their use in the field feasible.
- The more expensive units retain the automatic diaphragm operation of the lens.

Extension bellows consist of an accordion-pleated cloth tube mounted on a geared track. With a 50mm lens, bellows work best for subjects one inch high or smaller. They are larger and heavier than other devices but allow more flexibility for small subjects.

The advantages of extension bellows are:

- They are faster to adjust than extension tubes.
- The amount of adjustment is infinitely variable.
- Bellows permit greater image magnification than extension tubes or supplementary lenses.

Because extension devices increase the lens-to-film distance, the light intensity reaching the film decreases. To compensate for this, you must increase the exposure in proportion to the amount of extension used. Directions for calculating this exposure factor are supplied with the instructions for these devices. If your camera is equipped with a through-the-lens light meter, the meter will automatically compensate for the exposure factor.

CLOSE-UP TECHNIQUES

Camera steadiness becomes more important in close-up work, as any camera movement is magnified as much as the image is. A good sturdy tripod is recommended when it can be used. Unfortunately, most tripods can't be set low enough for many natural subjects. A bean bag, boat cushion, or rolled-up jacket can be a low-angle camera support. A cable release is also recommended.

Lighting is often a problem in the field, as many subjects are in the shade or in partial sunlight. A small electronic flash unit is a very handy accessory to supplement natural light. A few tests in close-up situations will help you determine the correct close-up exposure with flash. The short duration light output of the flash will also serve to freeze rapidly moving insects or wind-blown plants as well.

A *reflector* made of white posterboard or aluminum foil is also very useful for lightening shadows. Place the reflector as close as possible to the shaded side of the subject without getting it into the picture area. A reflector can also be used with a flash to lighten the harsh shadows created by direct flash.

Backgrounds can be a problem when the subject blends into the background so well that it becomes indistinct or when out-of-focus light areas distract from the subject. Artificial backgrounds can be made from cardboard or cloth. Natural colors, such as brown and green, tend to complement most natural subjects. Bright colors will give the photograph a studio appearance and may conflict with the colors in the subject. Keep the background distant enough to become slightly out of focus and free of shadows. Make sure the background is large enough to cover the entire picture area.

DOCUMENT COPYING

Effective lecture slides can often be produced by

photographing existing illustrations and graphics from textbooks, magazines, or other existing sources. A 35mm camera equipped with a close focusing device is ideal for this purpose.

Document copying is easiest if you mount the camera on a copystand. This is simply a vertical pole with a mounting plate for the camera (see Fig. 9-68). The height is adjustable, and the camera always stays parallel to the base. A photoflood light mounted on each side approximately 45 degrees above the base will provide glare-free illumination.

Fig. 9-69. Masking the camera with a hole cut in a piece of black cardboard will eliminate reflection of the camera in a piece of glass laid on top of the document being photographed.

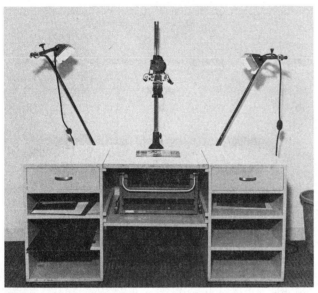

Fig. 9-68. A copystand is a device that holds a camera parallel to its base. A light on each side at about a 45-degree angle to the base insures even, reflection-free illumination for photographing flat documents.

Because good copystands are expensive, you may decide to build one. Plans to build a wooden model that works very well can be obtained from Eastman Kodak Company. Write to Eastman Kodak Company, Motion Picture and Audiovisual Markets Division, Rochester, New York 14650, and ask for Pamphlet No. T-43, *A Simple Wooden Copying Stand for Making Title Slides and Filmstrips.*

You can also use a tripod for copying, but tripods are awkward to use with the camera pointed straight down, and the legs often cast shadows on the material being photographed.

A sheet of ¼-inch plate glass is helpful to flatten documents that may be wrinkled or to hold magazine pages flat. Just place the glass over the document and shoot through it. To keep your camera from reflecting in the glass, mask the camera with a piece of black cardboard. Cut a hole in it for the camera lens and hold directly beneath the camera when you're making the exposure (see Fig. 9-69).

If you're copying material that's very light in tone, such as printed data on a white background, your camera's meter will read the white background as gray, and your slides will be underexposed. Make substitute readings off the palm of your hand or off a gray card rather than directly off the document.

WARNING: Most printed material is copyrighted and therefore illegal to reproduce.

◀ Planning a Slide Series ▶

Slides can help bring reality into a meeting or classroom. Slides are easy to organize and to keep up-to-date and convenient to store. Keep slide presentations fairly short—about 20 minutes maxi-

mum—and avoid too many slides with nothing but words.

There are many pitfalls in the hit-or-miss method of shooting slides first and then trying to assemble them into a story later. To become a good communicator with slides, you must plan in advance.

STORY-BOARDS GIVE RESULTS

The solution is to use a careful outline and then a story-board. With the story-board, you can organize your thinking to meet the needs of your audience.

The first step is to write one sentence that clearly states your purpose. This should be a very specific statement. For example, if you're preparing a slide series on extension work in your county, the statement, "To show extension work in Lea County," doesn't give you much direction. Instead, write a statement such as "How extension helps save dollars for our clients" or "How extension met the six needs pointed out by our advisory committee in 1981."

Next, write a general outline and decide about how many slides you want under each outline division. You may even want to jot down a short phrase to describe each slide.

Now you're ready to story board your slide series. An essential part of the story-board technique is the 3- × 5-inch story-board card. On the left side of the card is a rectangle in the shape of a 35mm slide (1 × 1½). Draw your picture ideas here. You don't need fancy artwork. Rough sketches will do the job.

At the right side of the rectangle, write a brief description of the illustration. With it you can successfully communicate your picture idea to your photographer or artist.

Next, write underneath the rectangle what you might say about the slide when it's on the screen.

ARRANGE CARDS INTO CHAPTERS

When you've filled out the cards, line them up in logical sequence. This arrangement, called the story-board, becomes the nucleus of your whole production. You can arrange the cards on a desk top or a wall board.

Usually the visualized story-board card will suggest natural chapters in the slide story. When needed, extra continuity cards visualizing the main chapter titles and summarizing the message can be placed in appropriate slots.

Now you can review your slide series and easily see weak spots that need help or areas that should be reshuffled. Ask your co-workers or clients to look over your story-board and make suggestions. This is the time to make changes.

The cards can be carried in your camera bag to remind you of the exact shot you need.

When all the slides are taken, you should be able to type your script directly from your cards with very

few changes. If you take the time for careful planning, your slide series will fulfill the goal you wrote in the first stage of making a slide series.

EFFECTIVE LOW-COST NEGATIVE SLIDES

While most of the slide stories you produce may be in color, there are a number of messages that you can present effectively by using low-cost negative slides. You can use this "quickie" technique with any illustrative material that is black on white. Some examples are charts or line drawings from publications, typewritten titles and test, and India ink drawings on white paper. You could prepare a complete slide presentation by using line drawings from a publication with appropriate typewritten titles. Such copy is readily available and easy to use. (See Chapter 10, for more details.)

Photograph the copy on high-contrast film, such as Kodalith ortho film, type 3, and process for maximum contrast. This film—the negative—is used as the actual slide.

To make these black-and-white negative slides blend with existing color transparencies in a slide presentation, color them. Use brightly colored pens on the emulsion side. Color individual lines on a chart. Different lines can be done in different colors. You can also use adhesive-backed colored film. This film should be the type used for overhead projection transparencies so the light can pass through it. This film, 3M, 15-1008-0, comes in a 20-sheet pack of rainbow colors. Simply cut the colored film to fit the area and place it on the back of the negative. Use a clean cloth to press out the air bubbles.

VERICOLOR SLIDE FILM

By using Kodak Vericolor SO-279 slide film and

Approximate Color Table for Kodak Vericolor Slide Film SO-279 with Black-on-White Artwork		
Desired Background Color		*Kodak Wratten Gelatin Filter*
Dark blue	12	(yellow)
Cyan	29	(red)
Green	34A	(deep magenta)
Red	38	(light blue)
Orange	44	(cyan)
Yellow	45	(deep blue)
Magenta	61	(deep green)
Yellow-brown	47	(deep blue)
Dark red		No filter

black type on white paper, you can make colored slides. They'll have various colored backgrounds with white or tinted letters.

To change the color of the background, place colored Wrattan gelatin filters in front of the lens. Experiment with exposure time, lens opening, and light source (either 3200° Kelvin Photo Flood or electronic flash).

Develop the film in standard Kodak C-41 chemistry and mount the slides. If you plan to use a great deal of film, buy a 100-foot roll of Vericolor 5072.

SLIDE FILING AND CARE

Slides are fragile. They can easily be damaged by fingerprints, scratches, fungus, or they can even be melted by a malfunctioning slide projector. Once you have put time and expense into producing some valuable slides, protect them by using a good storage and filing system.

Begin by duplicating any slides that are irreplaceable and use the duplicates for projection or to loan to others. Guard your original slides jealously. They are the film you exposed in the camera; there are no negatives to go back to.

If you're shooting a static subject and know that you'll need several copies of the slide, shoot several originals. This is cheaper than making duplicates, and the quality is better. Slides should be stored in transparent plastic sheets or in various types of file boxes (see Fig. 9-70). Keep slides in a cool, dry location, with a relative humidity of 60 percent or less.

Filing systems needn't be elaborate, but they should protect the slides from damage and allow rapid retrieval. One system that works well is a numerical structure based on primary categories and subcategories. Primary categories are given whole numbers, and subcategories are assigned decimal points under their whole numbers (see Fig. 9-71).

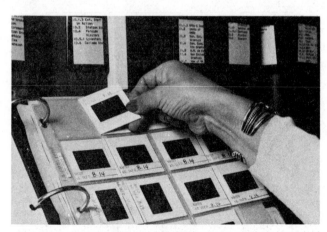

Fig. 9-70. Slides placed in slide-mount sheets inside a loose-leaf notebook make a nice slide storage and retrieval system.

The file number is written on the slide mount of each slide with permanent marking pen. This allows anyone to refile the slides once they have been pulled out for use. If a category gets too large, it can easily be broken into subcategories by adding decimal points to the number already on the slide mount.

Don't underestimate the value of a good slide col-

Example of Photo File

1.0 Agricultural Engineering
2.0 Agronomy
3.0 Livestock
4.0 Buildings
5.0 General Agriculture
6.0 Entomology
7.0 4-H
8.0 Home Economics
9.0 Horticulture
10.0 Individuals in College
11.0 Graphics
12.0 Poultry
13.0 Scenery
14.0 NMSU Programs
15.0 Extension Activities
16.0 Agricultural Economics
17.0 Agricultural Education
18.0 Agricultural Information
19.0 People in General
20.0 Wildlife and Fish
21.0 Branch Stations
22.0 Energy
23.0 Community Development

Expanded File System

3.0 Livestock

    3.1 Beef cattle

        3.1.1 Range
        3.1.2 Pen
        3.1.3 Equipment
        3.1.4 Sales
        3.1.5 Disease treatment
        3.1.6 Research

    3.2 Dairy cattle

        3.2.1 Equipment
        3.2.2 Production
        3.2.3 Disease
        3.2.4 Reproduction tests

NOTE: Each photo should have the file number on it so it can be refiled easily. File as few subjects as possible under the general category, such as 3.0, "Livestock." Instead, break the divisions into subdivisions. This system allows for expansion later simply by adding another number within a category.

Fig. 9-71. Examples of general categories and subject categories for a photo filing system.

lection. It can save hours or even days when you're producing a slide presentation. Seasonal events such as wheat harvesting are impossible to shoot during most of the year. Shoot them when they are happening, and with a good filing system, you will have the slides available when they're needed.

Bibliography

Cole, Stanley. *Amphoto Guide to Basic Photography*. Garden City, New York: Amphoto Books, 1979.

Editors of Eastman Kodak Company. *The Joy of Photography*. Rochester, New York: Eastman Kodak Company, 1979.

Hedgecoe, John. *The Photographer's Handbook*. New York: Alfred A. Knopf, Inc., 1977.

Lefkowitz, Lester. *The Manual of Close-up Photography*. Garden City, New York: Amphoto Books, 1979.

Nibbelink, Don D. *Picturing People*. Garden City, New York: Amphoto Books, 1976.

Rosen, Marvin J. *Introduction to Photography: A Self-directing Approach*. Boston: Houghton Mifflin Company, 1976.

Upton, Barbara, and John Upton. *Photography*. Boston: Little, Brown & Company, 1976.

Photo/Art Credits

Fritz Albert, The University of Wisconsin: Fig. 9-52b.

Michelle Bogre, USDA photo: Fig. 9-18.

Don Breneman, University of Minnesota: Figs. 9-23, 9-26, 9-27, 9-33, 9-42, 9-43, 9-48, 9-49, 9-63, 9-68, 9-69.

William E. Carnahan, Extension Service, USDA: Figs. 9-25, 9-30, 9-40, 9-46, 9-47, 9-52a.

Bob Coughlin, New Mexico State University: Figs. 9-7, 9-24, 9-50.

Victor Espinoza, New Mexico State University: Figs. 9-1, 9-8, 9-9, 9-10, 9-11, 9-12, 9-13, 9-14, 9-15, 9-16, 9-19, 9-20, 9-21, 9-22, 9-50, 9-51, 9-53, 9-54, 9-55, 9-56, 9-57, 9-58, 9-59, 9-60, 9-61, 9-62, 9-65, 9-67, 9-70.

Jeanne Gleason, New Mexico State University: Fig. 9-5.

Dave Hansen, University of Minnesota, USDA photos: Figs. 9-38, 9-39.

B. Wolfgang Hoffmann, The University of Wisconsin: Figs. 9-36, 9-66.

Karen Lilley, University of Minnesota: Figs. 9-3, 9-4.

Bill Marr, USDA photos: Figs. 9-29, 9-31, 9-44 (top), 9-45 (left).

Norman Newcomer, New Mexico State University: Fig. 9-6.

New York College of Agriculture and Life Sciences, Cornell University: Figs. 9-28, 9-32, 9-35, 9-41.

John Running, USDA photo: Fig. 9-37.

Bart Stewart, USDA photos: Figs. 9-44 (bottom), 9-45 (right).

John White, USDA photo: Fig. 9-34.

10   GRAPHICS

Graphics, like words, are a means of expression. Eloquence with either requires practice and good sense. Communicating through the visual elements of shapes, colors, and space is the subject of this chapter.

How much graphic production can you do? What difference can good graphics make toward the effectiveness of your newsletters, brochures, or visual aids?

This chapter is intended to equip you with the basics and a practical knowledge of the field, not to teach all the skills of a graphic designer. Even that much could make a great difference to your audiences. The more you use the principles of graphics, the more effective your communications will be.

Reader *acceptance* of printed material *is increased* when the material is attractively presented. Well-designed pieces are more *likely to be noticed, read, and remembered.* Messages are more *easily read* and concepts are more *quickly comprehended. Retention and recall are extended greatly* when the message is suitably illustrated.

Research supports each of these statements. If your aim is to communicate, you can't afford to ignore design principles. By definition, good graphic design is functional. It implies the planning and ordering of all visual elements. Communication is its basic aim.

In this chapter you'll learn many of the design devices used to direct the readers' attention. It's easy to begin to use them in your own communications, but beware of the pitfall called "overdone." Without realizing it, you can make things cluttered and ugly. Simplicity is still the best keyword to attractive design.

The total look of a communications piece conveys the total message. The color you choose adds meaning, or at least mood, to your message. The style of type you choose imparts meaning. And the way you put the message and graphics together will set a tone for the piece that may tell the audience "this is a dignified message" or "this is all in fun." In fact, your credibility could rest with the care you give to the total look of the communications piece. For example, which would most likely produce a good turnout for a meeting—typewritten on spirit master copies with poorly hand-drawn illustrations or a neatly typeset job with good photos and other graphics?

◀ Basic Design Principles ▶

Many of the underlying principles of design are traceable to the architectural efforts of the ancient Greeks. They *still* hold true. For instance, the proportion of an object isn't tied only to the rules of mathematics. You could measure to the center of a page, put your title words there, and still not be satisfied that the title looked centered.

Nor will it look pleasing. The human eye has its

Prepared by **Henri Drews,** extension information specialist, University of Minnesota.

own illusions. It perceives the center of a thing slightly above the measured center. We apply this knowledge by placing the words of a cover page above center so they "look right" (see Fig. 10-1). For another example, we make the bottom margin of a bordered area a bit wider to compensate for the same illusion.

Fig. 10-1. The illusion that center is higher than it in fact is makes this plan look "right." The bottom margin is wider, and attention is focused on the title above actual center. The readers' interest is increased because the words are dynamically balanced against the circular shape.

PROPORTION

Relationships of width to height are also based on ideals established by the Greeks. You may have noticed how the eye is attracted by regular shapes such as circles and squares. They hold our attention momentarily. Attractive, yet more interesting, is the shape of a rectangle.

The Greeks termed a rectangle 3 units by 5 units "the Golden Rectangle." They found that proportion to be the most agreeable to the eye. We still see these proportions today in 3 × 5 cards and 11 × 17 pads. The standard 8½ × 11 sheet, though not the same proportion, is nonetheless derived from the Greek's search for aesthetic perfection. The 8½ × 11 size is really a compromise with the economics of paper making. You will find that sort of compromise throughout the decisions you make related to graphics (see Fig. 10-2).

In addition to proportion, good graphic design relies on balance, contrast, and order to make a communications piece visually attractive.

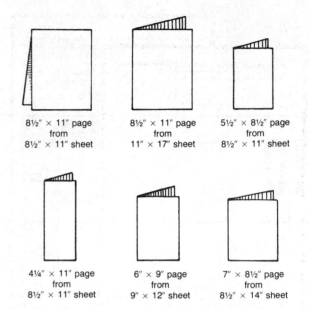

| 8½″ × 11″ page from 8½″ × 11″ sheet | 8½″ × 11″ page from 11″ × 17″ sheet | 5½″ × 8½″ page from 8½″ × 11″ sheet |
| 4¼″ × 11″ page from 8½″ × 11″ sheet | 6″ × 9″ page from 9″ × 12″ sheet | 7″ × 8½″ page from 8½″ × 14″ sheet |

Fig. 10-2. These formats are designed to make convenient-sized pages from common sheet sizes.

BALANCE, CONTRAST, ORDER

Balance need not mean having things equal on both the left and right sides of center (photos opposite photos or words opposite words). Rather, for the sake of visual interest, strive for dynamic balance. Balance the words in a block of copy against white space and a photograph. Think of three heavy people close to the center of a seesaw balanced by a thin person on the tip of the other end. Such balance creates the interest that makes the eye start in one place and want to go to the other. Fig. 10-3 shows examples of page layouts with dynamic balance.

Dividing the page into thirds or fifths is an easy way to assure that the layout will have interesting balance. When you begin to plan any on-going project, such as a newsletter, one of your first steps should be to decide on a format. Grids in divisions of three or five, such as those in Fig. 10-3, can serve as the basis for a newsletter format.

Contrast is a means of directing the readers' interest on the page. A dark place on a light page attracts interest, as does a light spot on a predominantly dark page. Large type amidst a page full of small type dominates the page. The same sort of emphasis is given to small type when it's contrasted against a surrounding of white space.

Order is a goal. There is nothing vague or random about good graphic design. Every element should be arranged to lead the eye progressively from one part to the next, enhancing the reader's comprehension until every part is seen. If any element distracts the reader or disrupts the orderly comprehension of the material, that is bad design.

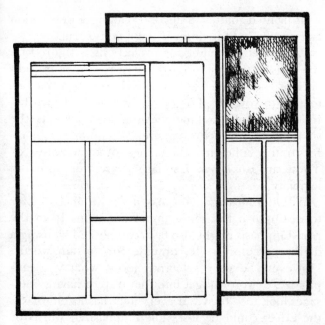

Fig. 10-3. The grid system divides the page both horizontally and vertically. Also, the shorter column width of these formats makes for easier reading.

COLOR

Choose *color* which is legible and which will fit the message psychologically. Select the dominant color to match the desired mood of the message. Red, for instance, fits a safety topic better than does purple. Blue is appropriate for a cold weather topic, etc. Some typical meanings that our culture assigns to colors are:

> White—fresh, pure, clean
> Black—dignified, strong, fearsome
> Red—beloved, angry, dangerous
> Blue—cool, melancholy, depressed, mildness
> Purple—rich, imperial, impassioned
> Orange—festive, urgent
> Yellow—warm, light, ripe
> Green—growing, youthful, sickly

Beware also of abusing the traditional combinations like the red and green of Christmas or the black and orange of Halloween.

Most of these color associations aren't new to you. We learn them through the messages we're exposed to every day on television and in magazines. Such daily indoctrination assures that your audience is sophisticated enough to tell when an inappropriate color scheme is chosen.

Usually for a pleasing arrangement you should use analogous colors—colors that are alike. Warm reds, oranges, and browns are alike. Cool blues and greens

are alike. If you imagine the colors of the spectrum bent into a wheel (so the infrared at one end meets the ultraviolet of the other end), analogous colors would be those which are adjacent.

Colors opposite each other on the color wheel are called complementary. Complementary colors are usually used as accents in a design scheme, if they're used at all. Schemes which use nearly equal parts of opposing colors are usually quite jarring to the eye. As a rule, avoid that type of situation.

The darkness or lightness of a color is very important when you consider the first criterion of color selection, legibility.

A tint of a color means white has been added to make the color lighter. Pink is a tint of red. A shade of a color means black has been added to make the color darker. Maroon is a shade of red.

You must make the words of the message legible. Choose the colors for the lettering combination with the background. Good combinations include black on yellow, black on white, blue on white, white on blue, or white on green. Poor choices include red on green, yellow on white, or black on deep purple. As a rule, the greater the contrast between the two colors, light to dark, the better the legibility will be.

Several hints about color use include:

● Select the appropriate dominant color first, then decide on a harmonious color scheme.

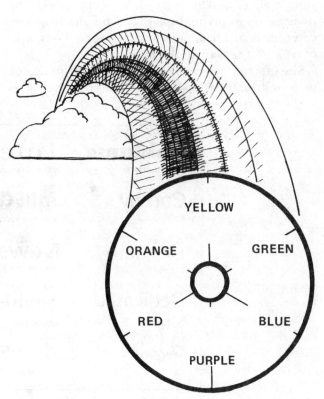

Fig. 10-4. On a color wheel, adjacent colors are called analogous. Opposite colors are called complementary.

- Keep the background neutral. You want the attention to focus on the message. "Cool" colors, because they recede, are generally more suitable for backgrounds, while "warm" colors, which advance, are usually better for the message portions, either graphic or verbal.

- For smaller areas, choose brighter colors. The brightest spot in a total scene will draw the readers' interest first.

TYPOGRAPHICS

The choices you make about type styles, sizes, case (capitalization), and spacing will also influence how your message is perceived. Look at the movie page of any newspaper to see how those four type characteristics can be manipulated to set the readers' concepts about movies. The meanings of the words are greatly enhanced by the right choices.

These next few paragraphs won't make you a good typographer, but they will alert you to some concepts you should know.

Type styles can be divided into two groups: *display* styles and *body* types. Display styles are larger and generally more ornate than body types. They're designed to enhance the meaning of words (see Fig. 10-5). The body types are simpler and more legible (see Fig. 10-6). The idea is to catch the attention of your readers and to set your concept quickly by using an appropriate display style for the headline, cover, or title. Then, use the more legible body style of type for the body of the message.

Size of lettering is more important to the design of projected material and display material than it is with printed matter. Someone reading a publication can usually hold it closer if the type is small.

Proper letter height for projections is figured on the basis of the distance to the last row of your audience or the distance at which you hope to attract attention to your display. Letters should be 1 inch tall for each 25 feet. That viewing distance sets the *minimum* for the lettering size. Minimum height refers to the "x height," the tallness of a lowercase "x." These are *minimums*. Use larger sizes for optimum legibility.

Within a message, you may use a variance of sizes to distinguish the more important items from the rest. Emphasis might also be accomplished by using a different color, style, or case for certain words. However, be very selective about what you emphasize, or you will get into that trap of having messages look too "busy." If every item has its emphasis, the effect could be lost. The emphasized words or phrases will practically cancel each other.

Just as style and color combinations can affect the legibility, so can the case. Use both uppercase and lowercase in your messages. Words in all uppercase letters are more difficult to read (see Fig. 10-7). One study notes a 16 percent loss in the speed of reading capitalized words. Such a time loss can seriously hamper audience comprehension in display situations.

Spacing also can be critical. Letters within a word should be close but not touching. Letters should look as though they have equal space between them, even if the spaces aren't really equal. For instance, some letters may tuck under other letters to look right. Trust your eye. Between words, leave the amount of space needed to insert a lowercase "i" comfortably.

Fig. 10-5. The style of type should fit the meanings of the words. In situations where readers must quickly catch the meaning, good typography is important. Type-style names are in small type above.

UNACCEPTABLE	ACCEPTABLE
Old English	Gothic Condensed
Brush	News Gothic
Script	Times Roman

Fig. 10-6. For the bulk of any message, choose type styles that are legible—such as Gothic or Roman. Gothic is characterized by uniformity of strokes and squared ends; whereas Roman is made up of thick and thin strokes with flared ends called "serifs."

ALL CAPS WORDS
ARE HARDER TO READ

because

they lack the unique
shapes of lowercase

Fig. 10-7. According to one study, the unique shape of lowercase words can add as much as 16 percent to the speed of recognition.

Lines of lettering should have enough space between them (leading) so that the tails (descenders) of lowercase letters don't touch the capitals in the lines below. Leading equal to one-half the height of the capital is usually sufficient, whether working with

one title or a whole paragraph. Incorrect use of space affects the continuity of reading.

Finally, don't stack letters and expect people to read them as words. English is most readable when printed or written horizontally, from left to right.

◀ Getting the Words Down ▶

There are a variety of ways to put words on paper or film for the sake of communication.

HAND LETTERING

Hand lettering might serve some of your needs, provided you understand the criteria for legibility, already mentioned, and exercise some skill and neatness. You'll get best results if you use a grid of horizontal and vertical lines (either on or under the

worksheet) to help keep the letters upright and straight. Plan the layout by lightly penciling in the message before you start to letter. The same precautions apply when using stencils, plus you must remember to fill in the gaps so the letters won't have that "fractured" look (see Fig. 10-9).

Wrong   Right

Fig. 10-9. Stenciled messages are more legible when you fill in the gaps.

Broad-point markers make it easy for you to produce your own hand-lettered signs and flipcharts. Several manufacturers make them with one-quarter-inch-wide or wider tips.

To learn to do a gothic-style alphabet suitable for most signs, make parallel lines in pencil on the paper to serve as guides. (A grid under the sheet will work as well.) Each letter will require several strokes, as shown in Fig. 10-10. Hold the marker so that only the beveled edge touches the paper. It should be turned in your hand so the edge marks its widest for all down strokes. For all across strokes, the marker should be rolled in your fingers 90 degrees, again so it makes its widest line. For round letters, use several overlapping strokes. With practice, you'll gain speed.

Gothic hand lettering isn't like writing longhand,

Fig. 10-8. With practice, you can adjust the spacing when you do the final lettering. The penciled plan just shows where you might increase the space between words to make them line up better. A soft eraser is a must to make the cleanup fast and easy.

ABCDEFGHIJKL
MNOPQRSTUV
WXYZ abcdefg
hijklmnopqrst
uvwxyz123456

Fig. 10-10. The beginner should practice the alphabet by using all the strokes. The more ornate styles of calligraphy require even more strokes per letter. With mastery, you can eliminate the strokes that you feel are unneeded, as long as you don't sacrifice good letter form.

because the wrist and fingers stay almost rigid. After learning the strokes well, go as fast as possible, and you will notice the letters will look more consistent.

The want ads of any newspaper, turned horizontally, are especially good for practice sheets (see Fig. 10-11). Ignore the words and use the columns as your line guides. Larger letters can be practiced by making them two or three columns tall.

As you gain proficiency, you may wish to use other tools such as lettering pens or brushes. The principles are the same. Styles are adjusted by holding the tools differently or by changing the number of strokes for each letter. For example, calligraphy for certificates is usually done with the pen tip held in the diagonal orientation (see Fig. 10-12), and fewer strokes are used per letter. Hand lettering stylebooks are available at bookstores and office supply stores.

Fig. 10-11. A newspaper's classified section, turned horizontally, provides an inexpensive, ready-made practice sheet for lettering exercises. To gain skill with large letters, make them two, three, or more columns tall.

Hand lettering never looks as professional as set type. It simply lacks uniformity and quality for the kind of messages that come from most information offices. Such professionalism is needed for credibility. But how can small budgets afford typeset jobs? We'll look at several alternatives.

Fig. 10-12. When lettering with a marker, you must hold the tip one of two ways, depending on the direction of the stroke you're making. The same is sometimes true of pen nibs, depending on the style of alphabet you want to make.

MACHINE-SET TYPE

First, keep in mind that the most common typesetting machine is the typewriter. It has drawbacks. The type is not as clean and consistent as professionally set type. Nor is there the wide choice of type styles from which to choose. Typewritten type isn't large enough for good headlines on a newsletter, much less large enough to use for overhead transparencies. Lastly, because of the unit spacing on most typewriters, it may be twice as wasteful of space as type set by other machines. You can fit the information from two single-spaced typewritten pages easily onto one typeset page.

Of course, the advantage of typewritten copy is the low cost. Many offices find that a good compromise is to have the typewritten body copy for their newsletters or brochures physically reduced 10 or 20 percent, using either a photostatic or an electrostatic copier (see Fig. 10-13). Then, they type the headlines normal size and paste it all together on their particular format with illustrations, etc. The results look neater than regular typing. That's one of those illu-

A column of type straight out of the typewriter is this size. It may be clean enough for reproduction in a newsletter but there are better alternatives. Compare this at 100% of the original size to . . .

● ● ●

This type from the same typewriter which was reduced to 90% of its original size. Besides the fact that you can get 10% more message into the same page, this type looks more professionally set. The copy equipment that does the reduction can also reduce to about . . .

● ● ●

80% of the original size. This is still quite legible for most readers, in fact since the eye will take in more words per eye fixation, this type should read slightly faster than the previous two examples. If you choose to use the white space that is gained as border, the psychological effect is to make the message seem more important.

Fig. 10-13. Standard office type looks cleaner and more suitable for use in a newsletter if it's reduced. The three examples shown above are one full size, one reduced to 90 percent of the original, and one reduced to 80 percent of the original.

sions caused by the reduction in size. Again, the reader can always hold the copy closer if it seems too small. However, with some audiences, such as the very young or the very old, good sense calls for larger type.

You may save by coordinating the job with your printer. He/she has to copy the material by camera anyway, so he/she could shoot all material one time and strip it together for you. Compare time and cost both ways.

Then there is rub-on type. If you need a small amount of professional-looking lettering, you should consider buying sheets of pressure-sensitive letters that are available at most office supply or art stores. They contain enough letters for multiple uses. They are available in several sizes and in every style. Again, with planning, care, and a minimum of skills or tools, you can execute the title or headlines you need.

Using rub-on type is slower than typewriting. It may take 15 minutes to do four or five words, but there are methods which can help. Here is a hint: Do all your words in rub-on type at once, giving all your concern to spacing and straightness. Then, you can cut them apart and paste them into your layout, aligned or centered with the rest of your message.

Fig. 10-14. As you transfer the letters, you should notice they change from black to gray. To avoid tearing letters, be sure they are all gray before you lift the transfer sheet away. Use light blue–penciled guide lines to keep everything straight.

The frequent need for many words in headline size (display) lettering may make it feasible for your office to buy a typesetting machine, such as Varityper or Kroy. The only disadvantage is the cost, because these machines are easy to operate and maintain. They are faster than pressure-sensitive letters but not quite up to typing speed. They provide good type of a size needed for titles, headlines, and overheads. Also, they allow flexibility of type style if you buy a modest collection of master disks.

The last alternative is to purchase professional typesetting services. This way you'll have the knowledge of the experts to aid your cause. Service is usually very fast. You can get type set even in small towns; you just have to know where to look. Any newspaper has the facilities to produce body copy and display type. If you ask for the service, they can supply it. Charges are reasonable. The type can be done to particular specifications of size, style, spacing, and column width, which can save you much preparation time. It's ready to be used in the galley form in which it's produced (see Fig. 10-15).

The galleys have several applications. They can be cut apart and pasted up for publications. They can be shot with a 35mm camera for title slides. They can be combined with drawings and run through a transparency processor to make excellent overheads. Once you get the type on paper, the communications piece is almost finished.

Fig. 10-15. A galley is a column of type as it comes from a professional typesetter. Keep it clean.

◀ Preparing Artwork for Publication ▶

Successful artists refine their illustrations from preliminary sketches and early drafts, just as good writers polish their work through several draft stages. To prepare artwork for a publication by any offset printer, you must also go through several stages. These are:

1. *Make a layout.* This is usually a pencil sketch. It shows where each headline, paragraph, and photo should fit. Care at this stage saves time and money later, even though you can never be totally accurate at figuring the size of paragraphs before the type is set.

2. *Mark up the manuscript.* In making the layout,
you decided which type should be enlarged, indented, or set in another style. Write instructions to the typesetter on the manuscript. If you're working with professional typesetters, they may assist you in the process. Just use the layout as your plan.

3. *Have the copy typeset.* Whatever your method of getting the type on paper, be sure it's as crisp, clean, and reproducible as possible. A second person should always proofread the copy to catch any errors. It's wise at this stage to make photocopies of the sheets of copy so it can be checked by all the individuals involved in the publication.

4. *Make a "dummy."* Use one of the typeset copies to put together a mock-up or dummy of the publication. This step is to see how everything will fit. You may not have the photos yet, but you should draw boxes where the photos will fit in the exact proportions you expect them to be.

The dummy may look rough because pieces will just be taped or glued in place, but it will show where paragraphs need to be shortened or where you must reduce a photo or even omit it. If the dummy is acceptable to all concerned, then go on.

5. *Prepare the paste-up.* This is the artwork, final and ready for the camera. At this stage, neatness and alignment count. Start by taping the paste-up sheet to the drawing board. Use the proper tools: T square, triangle, light blue pencil, ruling pens, etc. Next, draw the guidelines, sometimes called keylines, in light blue pencil. Light blue is used because the graphics cameras which are used to shoot the negatives for printing don't "see" light blue.

Draw any ink lines, dashed lines, or boxes now, because it will be more difficult after gluing or waxing steps.

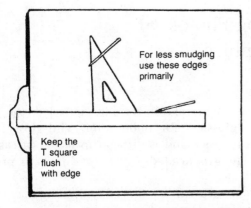

Fig. 10-16. The basic tools for preparing a paste-up are a drawing board, a T square, and a triangle. If kept clean and undamaged, they will last a long time.

6. *Paste the copy down.* This is actually part of the previous stage, but it deserves special attention. Use rubber cement, double-sided tape, glue, or wax on the back side of all the type and artwork (other than photos), then trim them closely with scissors. Neatly position the paragraphs with tweezers to avoid any smudging. Square things up perfectly, using the T square, then burnish them in place. Keep a protective sheet between the type you're burnishing and the actual rubbing tool, again to avoid smudging.

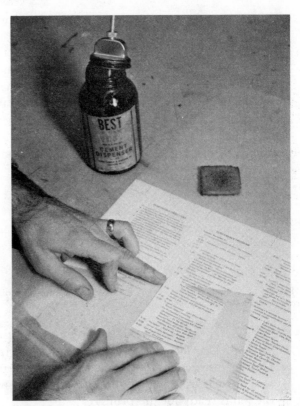

Fig. 10-17. Follow these tips to paste up with rubber cement.
a. Keyline the placement of every picture (or word block) with a light blue pencil.
b. Use cement that has been thinned to the consistency of cream.
c. Apply it to both surfaces and allow to dry.
d. Trim each item to ⅛ inch of the picture (or word block).
e. Lay a clean sheet of paper (preferably tracing paper) over all but a corner of the area where the item should go.
f. Position the item on the sheet of paper carefully, aligning it with the blue keylines.
g. Press together the cemented surfaces at the corners to hold the material in place; then remove the sheet from between the cemented areas and use it on top to protect the photo from fingerprints as you rub the surfaces tightly together.
h. Using a rubber cement "pick-up," clean the rubber cement from the edges.

7. *Apply photo windows or screened prints.* The selection, cropping, and scaling of photographs is covered later in this chapter. For this stage of the artwork preparation, you need to know that photos to be printed must be broken up into many minute dots of varying size. The amount of black ink and white paper in the printed photograph gives the illusion of a whole range of gray tones such as you would find in the original photo.

"Prescreened" photos can be applied directly on the paste-up. They are, however, quite coarse as compared to the "halftone" negatives your printer can provide.

If you have the printer make the halftones, all you need to do is put a red box exactly where a photo will appear on the paste-up.

When the negative of your paste-up is made, this red box will appear as a clear window where the printer will strip in the halftone of the appropriate photo. Red adhesive films are used to make the photo windows. Brand names include Rubylith, Zipatone, and others. Just cut and place carefully.

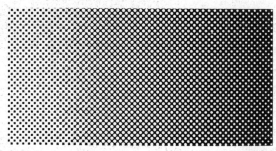

Fig. 10-18. A halftone photograph is made up of dots in various sizes. Look closely at any newspaper or magazine photo.

8. *Place any final instructions to the printer on the paste-up.* This includes any crop, fold, score, or register instructions that might be needed. Protect the artwork by giving it a cover flap. Now the job is ready for the printer.

Mimeographing is another method used for newsletters. Many small offices already have the machinery. They use secretarial time to type and produce publications. The advantages include low cost and complete control of the product. The trade-off is a loss of quality and limitations on the graphics you can use. For instance, unless you have an electronic stencil cutter, you can't use bold lettering, and you have to depend on someone's ability to trace drawings onto the stencil.

Without a stencil cutter, you must eliminate any plans to use photographs. Even with a stencil cutter, the photos must be done in a coarse-screened dot pattern, which can't show the detail possible in a halftone photo done by an offset printer.

The basic skills needed for mimeographing are simple to learn. They involve the use of a typewriter and a stylus to push wax around on the stencil. The ink flows through the stencil and onto each sheet of paper in the form of the words and drawings you want. Skills with stencils, bendays, mimeograph correction fluids, and styluses are best learned by instruction. Ask a mimeograph salesperson to help you.

◀ Selecting and Using Photos ▶
or Printed Graphics

We have all seen boring photos in print; the old "grip-and-grin" shot of someone receiving an award is a prime example of uninteresting, poor composition. (See Chapter 9 to learn how to overcome composition problems.)

What about the technical problems? We know a bad photo when we see one. The fuzzy focus and underexposure (or overexposure) give it away. If you want good graphics, you won't want either boring or poor quality photos in your publications. Don't accept inferior photography. It's usually worth the time it might take to get a better shot.

If the only shot you have is marginal and you must use it, talk to the printer who'll do the halftone. With special papers or careful exposures, he/she may be able to improve it, but don't expect miracles.

Chapter 9 describes how to take photos that are interesting, so here we will concentrate on selection and preparation of technically printable shots. Let your photoprocessor know you want *reproducible* prints, but not necessarily glossy finished. This may

surprise many old-timers, but you need not always have "glossies" for publication. The glossy finish tends to crack and is difficult to match in case any retouching is needed. *Unferrotyped finish* is preferable.

There are three criteria for judging a photo's printability: (1) good separation of adjacent tones, (2) sharp focus, and (3) wide range of grays (see Fig. 10-19).

Good separation of adjacent tones refers to the ability to distinguish the grays, whites, and blacks within the picture. It's important that the tones be clear because printing paper is seldom as white as the brightest spot on a photo and printing inks are never as black as the darkest areas of a photo. Therefore, when a picture is printed, the total range of grays is squeezed toward the middle.

Good separation of adjacent tones is particularly important in the photo's center of interest. Usually that means the faces of the people. If they're "washed out" by a flash that removes the tonal

Fig. 10-19. The photograph above (a) shows distinct separation of adjacent tones, (b) points up the sharpness of focus, and (c) indicates a full range of grays, from white to black.

"modeling," the faces will appear flat. The whole photo would have to be judged poor for lack of a good subject.

Sharp focus indicates the crispness of edges and detail in a photo. Fuzzy focus can't be improved in the printing process. The job is only made more difficult for the person who has to guess at the best focus to print a fuzzy negative. The focus is most critical at the center of interest. A photographer often will vignette the subject by leaving the background out of focus.

Wide range of grays means that the range between black and white in the photo should be complete. A photo with nothing lighter than middle gray tones will look underexposed or "muddy" when printed, whereas one with all light to middle tones is bound to appear overexposed. This range is important be-

cause it adds to the illusion of depth and contour we expect in a good photo.

CROPPING AND SCALING FOR THE PASTE-UP

The photo and its caption should tell a story. Between the two, nothing important should be left out. Too often too much information is left in. Verbose captions are an editorial problem. Cluttered pictures are a graphics problem.

"Cropping" means cutting the picture where you want. Unless you depend on a professional photographer, there will usually be unneeded background elements and other distractions that should be cropped from the picture. To emphasize the subject, you may crop the left edge of a photo and eliminate a distracting car bumper, for example. You may crop unsightly power lines by removing an inch from the top.

Whatever you do in cropping will affect the shape of the photo and how it will fit in your layout. You must keep in mind the proportions of the "window" on the artwork. That is the rectangular space you expect the picture to fit into when it is printed. Professionals use two L-shaped pieces of cardboard to decide how a photo should be cropped (see Fig. 10-20).

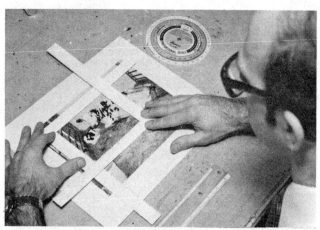

Fig. 10-20. L-shaped cards help in cropping by allowing you to see the picture without distracting portions. Keep in mind the proportions you want the final product to be—for you must maintain the same proportions in your cropping.

Rather than actually cutting the photo with scissors, it is best to indicate crop marks on the margin of the photo (see Fig. 10-21). If you carefully maintained the proportions of height to width, the photo should fit well on the printed piece.

Scaling is the process of sizing a photo for enlargement or reduction. A photo scaled at 200 per-

Fig. 10-21. The crop marks are made in red pencil or red ink on the margin of the photograph. Again, be sure they are parallel, square, and proportional.

cent will be printed twice its original size. Mathematically you can figure scaling as you would ratios:

$$\frac{\text{original height}}{\text{original width}} \times \frac{\text{printed height}}{\text{printed width}}$$

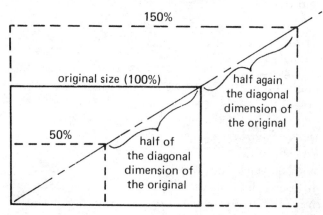

Fig. 10-22. Using the diagonal scaling method, you can figure the proportional enlargement or reduction of any picture. Draw a line through the corners diagonally. By doubling the length of the diagonal line, you will double the size of the picture, etc.

For example, assume you are starting with an 8 × 10 photograph, and you want to fill a space in a brochure column that is 3 inches wide and as much as 5 inches high. Which is the critical dimension? The height isn't as critical as the width because you could leave space at the top or bottom and the photo would still look right. So how would you crop and scale the photo?

Because 3 × 5 is proportionally taller than 8 × 10, you would look for portions at the sides of the photo that you can afford to crop. Let's assume the picture looked best by making it 4 inches wide. Use the ratio formula:

$$\frac{\substack{\text{cropped original}\\ \text{height?}}}{\substack{\text{cropped original}\\ \text{width 4''}}} \times \frac{\substack{\text{printed height}\\ \text{to be 5''}}}{\substack{\text{printed width}\\ \text{to be 3''}}} = \frac{20}{3x?} = \frac{6\frac{2}{3}}{1}$$

Therefore, you can mark the photo margins for cropping to a rectangle 4 × 6⅔ inches. Then, on a bit of paper taped to the picture edge, write a note to the printer explaining that you want the photo reduced from 4 × 6⅔ inches to fit the 3- × 5-inch space on the brochure panel. Construct your own scaling wheel—a tool to make it easy for anyone to figure reductions. (See the last page of this chapter.)

The printer is also the one you turn to when you want special effects done to the photo. If you need a wavy-line look to the whole picture, concentric circles, or some other texture, just ask. Perhaps you'd like to combine type with the photo or shape it differently; a good printer will tell you how or estimate the cost of doing it for you. As in any business, there can be a range of charges for a given job, so it's wise to do some shopping around. But when you find a printer you trust to give you a good value, cultivate good relations. Ask whenever you have a question and don't forget to praise a job well done.

CHARTS AND GRAPHS

A good graph or chart can say at a glance what might otherwise take paragraphs to explain. That's not to say words should be avoided, but rather that clear visual symbols, when combined with a few well-chosen words, make communication a painlessly efficient process.

Recognizing the strengths and weaknesses of the different charting methods can help you select the most effective one for your purpose and audience (see Fig. 10-23).

Are charts or graphs better than tables or text for expressing proportions, trends, quantities, and other data? Not necessarily in every instance, but when relationships of any concrete characteristics, such as size, position, temperature, quantity, and direction, must be expressed, a chart is probably the way to go. Abstract concepts, such as beauty, faith, and truth, are difficult to convey in visual terms (albeit not impossible). It depends on the information and its purpose. Use the following guidelines to best express yourself graphically.

● Use charts and graphs when words or tables are insufficient.

● Select the appropriate type of chart for the material and purpose.

● Consider the ultimate use of the information.

● Use recommended materials and reproduction techniques.

● Cultivate your "taste" for good design.

● Always follow the principles of legibility.

One study of how people prefer to get statistical information showed the first choice to be a well-designed graph supported by text. The second choice was a short summary table supplemented by text. The third choice was a graph alone. Fourth was a short summary table. It also showed that a good graph communicates more accurately than any other method studied.

Graphics emphasize important facts in the text. They substantiate by presenting the same material in different form, thus reinforcing the material in the readers' minds. To be most effective, they should be carefully integrated with the text that they summarize. A good graphic reveals at a glance relationships that might otherwise go unnoticed.

WHICH TYPES OF CHARTS TO USE?

To help you decide which types of charts fit your needs, the following is a review of some of the strong and weak points of the seven types of charts.

Bar charts are among the simplest yet most effective types. They work equally well in either horizontal or vertical form. This is useful because in publications, vertical bars usually fit the column format, while in audio-visual (A-V) presentations, projector screens usually fit horizontal layouts better. When horizontal bars are used, it's easier to place the labels on or near the bars.

	bar graph	pie chart	line graph	cosmo-graph	picto-graph	organi-zation	flow chart
Whole and its Parts	no	maybe	no	maybe	maybe	yes	maybe
Simple Comparisons	yes	yes	maybe	maybe	yes	maybe	no
Multiple Comparisons	yes	no	maybe	no	maybe	no	no
Trends	yes	no	yes	no	maybe	maybe	no
Frequencies	yes	no	yes	no	no	no	maybe
Sequences	no	no	maybe	maybe	no	yes	yes

Fig. 10-23. When is a particular type of visual best? This chart shows when one or more of seven types of visuals might be used effectively to illustrate a particular point or comparison.

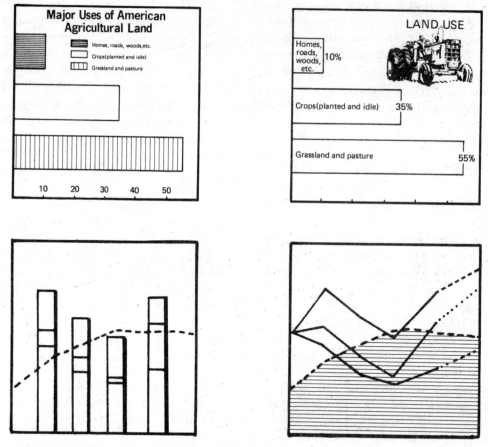

Fig. 10-24. Bar charts are versatile and easy to interpret if they are kept simple. Eliminate long titles, redundant scales, and complex legends, if simple labels will do.

Composed of measured and segmented bars, bar charts are commonly used with vertical and horizontal scales to compare different quantities at different times or under varying conditions (see Fig. 10-24). One bar with scale may be sufficient.

Pie charts are easily understood and are excellent for showing various parts that make up a whole (see Fig. 10-25). Thinking in terms of percents is helpful. For example, pie charts might portray the quantitative relationship of the types of farming that make up the state's agriculture or a breakdown of the average family's food dollar. Each segment of the chart should be clearly defined in line or color and have a label on or near it.

Be aware that unnecessarily adding the dimension of depth to a pie chart may confuse readers.

It's not always necessary to complete all 360 degrees of a pie chart. If labeling only portion(s) does the job, don't be concerned with the remainder. Think in terms that are significant to your audience.

Line graphs are the most common type. They are

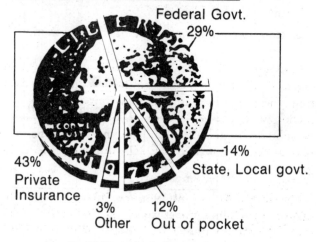

Fig. 10-25. A simple illustrated pie chart.

particularly useful for showing trends or comparing relationships. A single line can be used to indicate growth or expansion. A few different (for example, solid and dotted) lines might compare related items, such as salaries, with the cost of living. But when the readers should note specific quantities, a bar graph is more efficient.

More than three lines on one chart often causes confusion and even misinterpretation by the viewers (see Fig. 10-26). So it's preferable to break down the information into a series of simple charts and present them step by step. Experiments with grouped line charts demonstrate that even simple line intersections confuse readers.

Fig. 10-27). For example, a pictograph of the "Growth of Retail Food Outlets" might use stacks of coin symbols as bars and include a drawing of a person at a checkout counter somewhere on the chart. A variety of common symbols is available commercially in rub-on sheets, and many others can be found in clip art sources, such as ACE's *Clip Art Book Number 5,* available from The Interstate Printers & Publishers, Inc., Jackson at Van Buren, Danville, Illinois 61832-0594.

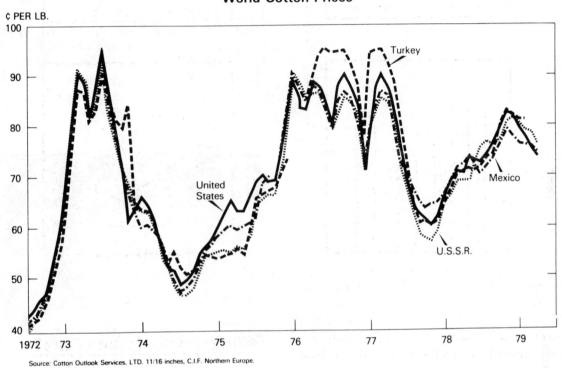

World Cotton Prices

Source: Cotton Outlook Services, LTD. 11/16 inches, C.I.F. Northern Europe.

Fig. 10-26. Multiple lines on a chart can be confusing. Use caution.

Cosmographs, like pie charts, mainly depict a whole object divided into parts. Another element, such as time, location, or value, is usually implied by the direction or structure of the chart. Cosmographs could be used to represent assemblies, for example, how part of one organization also fits into a different organization.

Pictographs use simple visual symbols to indicate quantities. A silhouette of a car may indicate 100 actual cars, or an image of a house may represent 1,000 houses. Symbols in this case contribute to quicker and sometimes more accurate interpretation of the material by the audience, especially when many different topics are combined.

The Graphics Institute of New York recommends combining a "sign post" drawing with pictograph symbols for maximum effectiveness and interest (see

Organization charts show how parts of an entity fit the hierarchy of the whole. For example, a company's organization shows how personnel fit into the corporate structure. A series of connecting lines depicts the chain of authority and working relationships. Organization charts are excellent for explaining the staffing and functions of any enterprise.

Careful planning must be done to indicate correct relationships, level for level, and to insure legibility of the finished product. Since personnel are bound to change, it's wise to keep a copy of a good organizational chart on file for updating as needed.

Flow charts present sequences in time or progressions and are useful for showing everything, from the procedures for mixing and applying herbicides to the intricacies of the U.S. educational system. These charts are often combined with text, using ar-

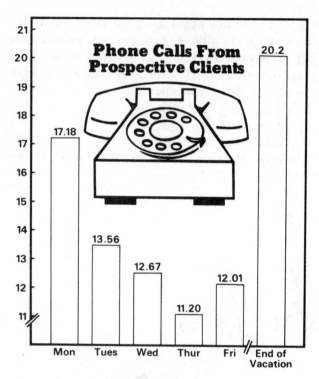

Fig. 10-27. A good "sign post" drawing lets the readers know what the chart is about, even before they read the title.

rows, numbers, or other symbols to show the progression from one step to the next.

Combinations may be the answer when special interpretations of data necessitate the use of more than two elements common to most charts and graphs. For example, combining a bar chart with a line chart can present elements of time, amount, and price. Three-dimensional graphics are difficult for the average person to interpret, but they may be used for a group of scientists, for example.

Design charts to be as simple or detailed as suits your audience. To be sure the final result will be interpreted accurately, always test any combination of charts with a sample of the intended audience while the graphic is still in the layout stage of development.

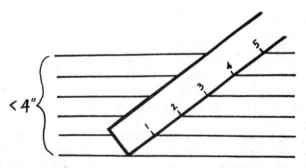

Fig. 10-28. A quick way to divide the space for a graph into equal segments is to place a ruler diagonally across the allotted space and then use any number that fits your purpose.

CHARTING HINTS

Borrow ideas from the professionals. *U.S. News & World Report, Time* magazine, and similar publications make very effective use of charts and graphs. Refer to them occasionally. Keep a file of any ideas and techniques you might be able to use in your work. You can purchase most tools and supplies directly from the manufacturer or local art and office supply stores.

Use the right tools (see Fig. 10-29). Since crisp, solid lines are a must for good reproductions, prepare your originals using black drawing ink (India ink) on good-quality, smooth-surfaced paper or illustration board. On these surfaces, you can make corrections by erasing, scraping with a sharp blade, painting over with opaque white water-base paint, or carefully pasting a new piece of paper over the mistake.

Fig. 10-29. Tools needed to make good, clean graphics.

For inking lines, use a ruling pen, which is inexpensive yet adjusts to many line widths. You may prefer a set of technical pens. They have points of varying widths for drawing heavy or light lines uniformly.

Other tools you'll need include a T square to insure that everything begins and continues straight, masking tape to hold the paper secure, a 30°-60°-90° triangle to align things vertically, an inking compass or circle template for circles, a French curve for irregular curves, and possibly a protractor.

Measurements for plotting your graphs can be done with a ruler. And, as in the layout for publications, you should use a light blue pencil to plan every line and label before inking. It helps to get everything just right.

If you use colored papers and tapes for your

charts, you will also need an art knife. Buy the kind with changeable blades to assure a sharp, clean cut every time.

It helps to write a descriptive paragraph before beginning your chart layout. This organizes in your mind the points that need visualization. In visualizing, think of the associations your viewers might look for, or create new ones. Start with your primary audience and purpose in mind, but don't overlook possible alternative uses for the material. Will the published study be reported at a future meeting? If so, you should plan originals that can be photographed for slides or made into overhead projectuals without redoing the artwork. Considerable time and money can be saved with such foresight.

Any illustrative material that aids viewers in the correct interpretation of your graphic is probably worth the trouble. In addition to "sign post" drawings and symbols, cartoons and photographs can make data more palatable to your audience. Photographs can be used as bars or segments of pie charts or as backgrounds for charts. When used as backgrounds on which printing will be superimposed, the photographs should be light and have little contrast.

Color makes for effective graphics, especially those that are to be projected. It can make a visual more attractive, differentiate between items, provide several tones through screening, and aid associations to realism—green for pasture acreage, gold for wheat harvested, and so forth.

Don't crowd elements—leave white space for easy reading. If your graphic is to be printed, be sure the artwork and lettering will reduce and still be legible. Make the original larger than the size of the reproduction in the publication. The larger size is easier to work on and imperfections in the drawing will be minimized in the final reductions.

Eliminate all unnecessary wording in the layout stage, and avoid vertical lettering. Stacked letters are very hard to read.

Stacking words
is all right,
but don't stack l
       e
       t
       t
       e
       r
       s

If someone else will be doing the final artwork, provide the artist with a carefully drawn layout. Type or print all information and instructions. Carelessly scribbled layouts only cause mistakes and

waste. Be available if the artist has questions. Check for errors before the art is in the final stages. Allow for suggestions from artists. Creativity is their business. But judge the product ultimately against your knowledge of your audience.

ILLUSTRATIONS

Illustrations can be the essence of your communication piece or merely cosmetic additions. Both uses are valid. Style and quality are important in either case because, like other graphic elements, the illustrations you choose strongly influence the viewers' perceptions of the message (see Fig. 10-30). You wouldn't put a poorly drawn heart or liver with a medical journal article, would you?

"When men are unfairly treated they naturally turn to desperate courses." Procopius AD551

Fig. 10-30. The illustration strongly influences the viewers' perception of the message. Choose carefully.

You should use drawings that are contemporary by their style. Dated drawings will make your information appear outdated as well. Learn to recognize by the linework the differences that make a political cartoonist's rendition insulting, while a portrait artist's lines may be flattering, although there are times when one uses the other's technique to suit a situation. Good illustrations have a lively quality that makes them believable. As with your photos, try to use only the best.

For most applications of art in publications and projected visuals, it's usually less expensive and more effective to use line art rather than tone art. A black-and-white photographic print would be an example of tone art. Other examples include most charcoal, pencil, and wash drawings. They all have varying tones of gray to suggest the structure of the objects in the picture. Line art, on the other hand, is comprised only of black and white, no grays. For instance, an ink drawing has areas of white with lines of black and may use "hatched" lines to suggest tonal changes. Screened photographs are really a form of line art.

The illustrator can emphasize special features of an object through exaggeration. For instance, the vein pattern of a leaf is more easily seen in a drawing than in a photo. The illustrator can "explode" a mechanical assembly to show how the components fit together. He/she can indicate impressions, such as motion and heat radiation, that the camera has difficulty showing.

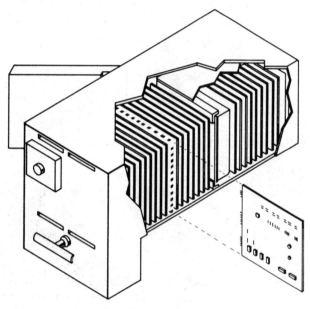

Fig. 10-31. Line illustrations easily show views which would be impossible or expensive for a photograph.

If the subject is familiar, cartoons are as easily understood as realistic pictures. The humorous approach is usually more attractive, but satire, irony, and allegory in cartoons can be misunderstood, so use them carefully. Choose humor that is clear and inoffensive.

The illustrator can suit your message and audience without leaving the studio. Unlike a photographer, the illustrator doesn't have to rely on things that exist. He/she can draw situations that have never happened except in the imagination.

Illustrators today have found, however, that their time is spent more efficiently if they use some of the technology of the photographer. The techniques are simple enough to be used by teachers and other communicators on a small scale with just a little practice. One technique is called "photosketching." Basi-

cally, it entails tracing from a photograph or printed picture. The advantages of doing so may not all be obvious.

- The product is immediately ready for reproduction by almost any method.
- Photosketching is faster than drawing freehand because it eliminates the proportional structuring steps—the rough sketches.
- All distractions are eliminated from the photo.
- Images can be combined with those of other photos or with words, symbols, maps, etc.
- Phototracing is *not* an infringement of copyright laws.

To photosketch, start with an idea and then search for pictures that relate to that idea (see Fig. 10-32). Select the portions of one or two that convey your idea. Trace them onto a sheet of paper. The drawing can be used as a master for overheads or artwork for printing.

TIPS ON PHOTOSKETCHING

To make photosketching easy, start collecting a file of sources. Save trade journals that relate to your field of expertise. For instance, if you're an extension home economist, your reading material includes a variety of magazines related to your field. Save pictures or whole issues. Even if you have a good memory, it's a good idea to categorize the topics. A source of general items such as a mail order catalogue is valuable too, not only for objects but also for figures and gestures.

Tracing is easiest when you use a good drafting

Fig. 10-32. Using the photosketch method, you can construct complex illustrations by combining various images, such as those found in a catalogue. Be careful to select images that are in scale with each other.

vellum. It's translucent and tough enough to stand repeated erasures. You may get it in any office supply store.

India ink and a drawing pen or fine brush are the preferred utensils, but felt-tip pens may be less fuss. If you intend to make an overhead of the photosketch, be aware that most felt-tip penlines won't reproduce on a standard office thermographic copier.

The image of the picture is more easily seen to be traced if the paper is taped atop a light box. You can use a window also. Print and pictures from the other side of the page sometimes confuse the image.

With practice, you'll learn not to overdo the line work. You'll gain proficiency at shading and drawing style. Often you'll need to transfer a line drawing to another surface. A good alternative to carbon paper is pencil shading on the backside of the drawing. Lay the drawing over the new surface and go over the lines again with a No. 3 pencil to transfer your artwork.

USING CLIP ART

Clip art refers to drawings done by professionals that are copied and sold for reuse. When you buy clip art, you also buy the copyright, so it can be used as you see fit. Clip art is usually inexpensive, when compared with the cost of a professional artist or the time it takes to do your own artwork. The problem is that the picture usually isn't quite what you had in mind. Clip art is designed to fit a wide variety of uses. It's categorized in groupings, such as business, sales, seasons, travel, etc. Commercial clip art services include Volk Corporation, Dover Publications, Dynamic Graphics, and others.

Special mention should be made for the other major ACE publication—*Clip Art Book Number 5*. It features art of most agricultural and rural life topics at a price lower than many other sources.

Most clip art is printed in two or more sizes because pictures for projected visuals must be larger than those for publications. You'd be wise, however, to preserve your file of art by using only copies of the pictures you want. If you use photostatic or electrostatic copies, you can adjust the size of the art to your particular needs while preserving the original for future uses.

Sometimes a piece of clip art isn't quite what you want. If so, change it (see Fig. 10-33). With skill and care you can change the length of hair in a picture, put something in a hand, change a gesture, remove something, add a woman to a scene that had all men. You may find the ideal image is a combination of two or more bits of clip art. Just paste things neatly in place and, voilá—you're an artist!

Fig. 10-33. You can change clip art if it seems out-of-date or not quite what you need. Erase the unwanted lines or use typing correction fluid to cover them. Make the changes with black felt-tip pen or India ink to get just the effect you want.

◀ Visuals for Meetings or Classes ▶

Why use visuals? They increase interest and attention given to the program. But, mainly, they help your audience to remember what you say. In one study, the same material was taught to two audiences. One lecture was with visual support, the other, without. Short-term recall of the material was affected somewhat according to a test three hours after the lecture. The audience shown visuals scored 85 percent versus 70 percent for the other audience.

The dramatic effect was in long-term recall. Ten days later, the group that had seen visuals recalled 65 percent of the material, while the group that hadn't could recall only 10 percent.

People are very dependent on their vision. It's estimated that 83 percent of all we learn comes through our sense of sight. What you learn about graphics can be applied to visuals to accomplish several goals simultaneously.

If the audience is receptive, learning is easier and seems more enjoyable. Education is akin to entertainment, when you use visuals. They keep your viewers' attention focused, by keeping their eyes on your topic and away from any distractions.

Concepts are understood faster. Visuals provide a setting for more specific information on the topic shown. Understanding difficult concepts such as those about foreign places and new experiences is made easier with visuals. Progress is logical; the planning of a visual talk assures better order. Your audience can see where it's going. Misconceptions are avoided. Good visuals clarify, illustrate, and support the verbal material. Information is retained. The message—the reason for your talk—will be remembered longer and by more people.

Visible
Interesting
Structured
Useful
Accurate
Legitimate
Simple

Fig. 10-34. The qualities essential to good visuals are so important that the rule against stacking letters is broken here to help you remember them.

Which method is appropriate? You could choose to support your talk with a chalkboard, flipcharts, Velcro or magnet boards, flannel graphs, models, actual objects, overhead transparencies, slides, or filmstrips. All these methods rely on short, concise messages. Use handouts if the material is long or so important that the members of the audience should have a copy. Consider that each method has strengths and weaknesses. Then, choose the method that is most legible for the audience and most comfortable for you, the presenter.

Visuals fall into two groups—nonprojected visuals and projected visuals. Learning is enhanced if you as the presenter interact with the visual aid. Your activity with the visual, building the story step by step, serves to involve the audience. It gives the impression that the presentation is specifically for them. Because the presenter is always in front of them and because even simple devices sometimes don't work, these active visual methods require that you develop the ability to ad-lib a bit more than when using packaged projected visuals or no visuals at all.

CHALKBOARDS

Chalkboards are the most common visual device. With proper use, they can be quite effective. They don't require preparation, although preparing important portions prior to your talk may be a good idea. For example, tables and literary excerpts take valuable time to write and should be checked for errors. Learn to print neatly and large enough for the back row sitters—about 3 inches high. Use the top of the board so all can see. Colored chalks are available for emphasis and definition of materials. In fact, for use on TV, yellow chalk on green board is recommended. It reduces the glare problem.

Paint stores carry chalkboard "slating" paint, in case you should ever want to turn any smooth surface into a chalkboard. A disadvantage: Material done in chalk must be repeated for each talk, which may suit math professors but seems inefficient for other people.

FLIPCHARTS

Flipcharts are active visuals that can be effectively used in teaching. You may create them ahead of time or during the presentation. You may emphasize a specific idea by pointing at the illustration or copy, or by drawing or writing some additional information on a new sheet. Material may easily be reviewed by turning back to visuals used earlier.

Paper easel pads are commonly used for chart presentations. Many folding aluminum easels are very portable and convenient. The most available pad size is 34 inches high × 27 inches wide. Some easels allow the pad to be used horizontally. Since we write horizontally, this accommodates more letters per line. Many chart users prefer the easel that uses the 34- × 34-inch pad. Some easels have legs that are adjustable in height, which allows some flexibility to maintain visibility in varying situations.

Easel pads are available in a variety of paper types. Among the most popular are newsprint, most inexpensive but poor quality paper; ruled, a better quality white paper with light blue lines ruled every inch; and blue, which reduces contrast and is more acceptable for use on television.

Fig. 10-35. You can draw different pictures lightly with pencil ahead of presentation time. During the presentation, you can use colored chalk, wax crayons, or wide-point felt-tip markers.

Wide-point felt-tips markers are often used on easel pads, but they bleed through onto the next page. A paper slipsheet may be used under the page upon which you are lettering to avoid bleed through.

Wax crayons don't bleed through, and they are available in sizes thick enough to create good heavy lines that are visible at a distance.

Oil pastels are preferred by some chartmakers because they are available in a wide variety of colors and in many thicknesses, and because they don't bleed through, but yield a good, dense layer of color. You must grip the sticks close to the paper so they don't break.

Charts are often prepared on posterboard, which is available in an assortment of colors and is usually 22 inches × 28 inches in size. It's often convenient to place these on an easel during the presentation. Felt-tip pens work well on the slick surface of posterboard and, because of its thickness, the ink doesn't bleed through.

VELCRO BOARDS, MAGNET BOARDS, AND FLANNEL GRAPHS

The use, merits, and problems of these three types of active visuals are similar. They differ only in the ways they work. Velcro depends on the hook-'n-loop

material available in fabric stores. Magnet boards require thin sheet metal and small magnets or magnetic tape from hardware stores. Flannel graphs use a special velvet-like fabric to make the symbols stick to flannel-covered board. Many are designed to fold for storage.

The greatest advantage of Velcro boards, magnet boards, and flannel graphs is that you have total control of the timing and presentation techniques. These items are portable, durable, and colorful. Using markers or tempera paints, you may prepare the message on posterboard cut-outs.

Making the shape of each piece of the message agree with the words is effective. Incorporate light objects such as Styrofoam and yarn for interest. Magazines are a rich source of illustrations. Just cut out a picture, and paste it on a piece of board. Glue the Velcro, magnet, or velvet paper on the back, and it's ready. Keep the colors bright and provide good contrast for the lettering to be read.

When using these tools, arrange the pieces in order of use and handy to the board. Place them quickly but carefully, and face your audience as much as possible. Try to avoid doing any of the following.

- *Symbol-waving.* The speaker jabs at the audience with the visual as his/her weapon. He/she may also tease the audience with subtle glimpses of the words to come or give those watching a clinical view of how the backing is placed on the cards.

- *Discarding.* The speaker brings more symbols than can be used. He/she leans over the stack, studies the next symbol briefly, decides not to use it, tosses it aside, studies the next, and so on. The audience is more intrigued by the unseen than by the seen.

- *Anticipating.* The speaker's hand anticipates his/her voice. The instant the symbol is revealed, every mind in the audience registers its meaning. Not only does the audience miss what the speaker is saying, but it has already lost interest in what he/she is about to say.

- *Troweling.* The speaker places each symbol carefully, then fondly smoothes it and flattens it with the gesture of a crafts worker using a trowel to finish a freshly laid batch of concrete. The audience's attention is focused on the pantomime, not the message.

- *Backsiding.* The speaker turns away from the audience as if more fascinated by the artistry of his/her handiwork than by the group he/she is trying to reach. Nobody can arrange symbols properly without momentarily turning toward the board. But the loss of eye contact with the audience should be so momentary as to go almost unnoticed.

● **Blocking.** The speaker blocks all or part of the view of the board. Here again, momentary blocking is unavoidable. If the speaker really wants the audience to forget the visuals for a moment while he/she makes a point, the block can be used effectively. When placing visuals on the board, he/she should stand aside so that as little of the audience's view as possible is blocked.

● **Telescoping.** The speaker places a series of symbols on the board in such rapid sequence that the audience doesn't have ample time to recognize and understand the meaning of each. Tip: The boards can be photographed step by step and made into a set of slides.

MODELS

Models can be effective in teaching structures and processes to small groups. They don't lend themselves well to large groups because if they were large enough to be seen, they would be too bulky to handle. Models can be made of almost anything.

Fig. 10-36. This model of a house roof was made with Fome-core (trade name) and masking tape. Models serve only small groups well.

You'd do well, though, to borrow an idea from architecture students: Use balsa wood and a new type of board material made of polystyrene sandwiched between paper. These materials are easily pinned, glued, or taped together.

ACTUAL OBJECTS

There is nothing like the real thing for a learning experience. Whether it's a piece of machinery or a stuffed animal, the audience can get concrete ideas from seeing and touching the real thing. Pass it around if possible.

Realia won't work well for a large audience. Of course, extremely large or extremely small objects won't work either.

PROJECTED VISUALS

It's advisable to use some form of projection when you have an audience of 100 or more. It will be difficult for audiences of that size to see charts, posters, and models, even if lighting, seating, color combinations, and other physical conditions are just right. A visual that can't be seen is worse than no visual at all.

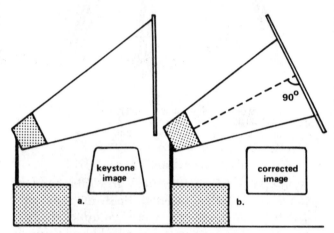

Fig. 10-37. Solve keystone distortion by making the screen perpendicular to the light source.

Usually, you have little control over the room in which you use projection, but whenever possible, strive for the preferred arrangement: The first row of seats no closer than 2½ times the width of the screen being used, the last row no further than 6 times the width, and no one beyond 30 degrees from the line of projection, unless a lenticular (wide-angle) screen is used.

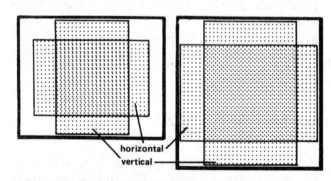

Fig. 10-38. Check the fit of the images on the screen beforehand to be sure you won't project on the wall or ceiling.

VISUALIZATION

Visuals should be more than just projected words for the presenter to read to the audience. In fact, the audience can read them faster. To remember the idea, viewers must be able to visualize it; that is, they must be able to put it into a pictorial context. Good students in the audience often do the visualization internally. A good presenter will help the whole audience by making the visual associations for them. That's all a picture is when it's used in a visual aid—an association.

Psychologists have lists of strongly associated words, such as "light/dark," "table/chair," "fork/spoon." Similarly, when you're trying to visualize an idea for a talk, ask yourself what other things, places, or people you associate with it. Some visualizations are better than others, of course, depending on their strength of association and the dramatic impact of their composition.

FORMATS

Plan the words and pictures of your visual to fit in a horizontal format. This best fits most projection screens (including television screens), so you can fill the width of the screen with image. For overheads, the proportions of the format are usually 4:3. For 35mm slides, the proportions are 3:2. Use multiples of those ratios for doing the actual artwork.

◀ Overhead Transparencies and Slides ▶

ADVANTAGES OF OVERHEADS

Overhead transparencies are perhaps the most versatile visuals for meetings. They can be carried in a briefcase. The projector is practically a standard piece of equipment at any regular meeting place. The room need not be darkened. Dim lights may be on for those in the audience who take notes. Any of the buildup or discovery techniques of the active graphics boards can be done on an overhead. Silhouettes of actual items can be shown if they are small enough to fit on the projector stage. The transparency can be used as a chalkboard. The presenter has the advantage of writing in normal handwriting size and position without turning away from the audience. Various colored pens and markers are available to spice up the presentation. Clean up is easier than on the chalkboard also; simply put a new sheet of acetate in place of the used one. Unlike with a slide set, the presenter has the ability to rearrange the order of the visuals without difficulty.

Any image that can be put on slides can also be made into an overhead, although photographic images are considerably more expensive in this large size.

HOW TO MAKE OVERHEAD TRANSPARENCIES

Transparencies can be made in at least six different ways: printing, photography, diazo, electrostatic copier, thermographic copier, or by hand.

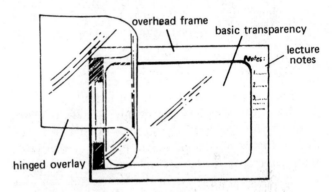

Fig. 10-39. Tips for the overhead user include:
a. Taping the basic transparency to the back of the frame all around so no edges can catch when you file the transparency.
b. Writing lecture notes on the frame.
c. Using masks and overlays to help explain your material, but planning how they will hinge so they won't be hard to use, won't be distracting, and won't interfere with the projector lens column.

The *printing* method is used by commercial producers of audio-visual products. Printing transparencies are reproduced on huge presses in large quantities. This is not likely to be the method you use.

Photographic transparencies are simply enlargements on film. Therefore, any photoprocessor can make them for you. Ask for an 8×10 print on film. Be sure to say, "Lightly print it for use on an overhead projector." Cost and quality will both be high compared to other methods of making transparencies. Smaller black-and-white prints on film can be taped into or onto overheads to combine successfully, for instance, a person's portrait with a quotation lettered on the acetate. This can even be done with Polaroid photos, if you use the right film.

Diazo transparencies are inexpensive and almost as faithful in detail as photography. They are available in a wide range of colors on clear as well as black on tinted acetate. The diazo "dyes" are burned out by exposure to ultraviolet light, then ammonia fumes develop the image in a color on the acetate. To have colors, you must sandwich two acetates. If you don't have the diazo machinery, the copying service can usually be bought from any business that does blueprinting. However, the master copy must be translucent, which means original artwork should be done in black ink on vellum or acetate. That can be a problem for offices not equipped to put type on those surfaces.

Electrostatic is the generic term for the Xerox process. Transparencies are possible with this process if you load the machine with acetate sheets made for such use. The artwork must be fine line. Bold areas tend not to reproduce well. The surface is also difficult to work with if you intend to enhance the image using tapes, markers, or appliques. New models of electrostatic copiers can copy in color; even local printers may offer that service.

The *thermographic* process is common in most offices. It is a process which copies carbon black such as black ink, print, graphite, etc., using infrared light to darken the acetate by heat. The acetate is made especially for this use. The process doesn't copy the lines of most marker pens. However, since electrostatic paper copies make excellent masters, you can use the Xerox copier to make the master for this process.

The last method of getting lines and lettering on overheads is *by hand*. Basically, the best advice is to trace carefully the images and lettering you need. The major drawback is that such overheads are slow to prepare. Also, should you need duplicates of the transparencies, there's no quick way to get them.

POOR	GOOD
elite executive pica	Directory Speech-Riter OLYMPIA

Fig. 10-40. Standard office type is less than ¼ inch tall; therefore, it is unacceptable for use on overheads. However, some special typewriters with larger type can be used.

The materials used to enhance the overheads with color and borders are the same for any of the methods you may choose. The trick is to use them to aid the message.

Grease pencils and marking pens are available in several colors. Some are permanent, some washable.

Transparent tapes come in several colors and widths from $1/16$ inch to 1 inch. They are good for making quick bar charts, trend charts, and borders or underlines. When joining lines with tape, overlap the ends, then cut carefully across the intersection with a sharp blade. Peel away the excess and you have a very neat splice.

Color applique: Color adhesive acetate sheets for use on overheads are available also. These are used to fill large areas with color. Pockets of air tend to get caught between the applique and the acetate, unless the applique is laid very carefully. You can cut the applique after you have applied it to the acetate, but don't cut through both sheets. Practice to develop the touch. The air pockets can be pierced and deflated to make them less noticeable.

Special inks are also made to color transparencies. They can be blended, but beware of puddling.

Mount overheads on frames to make them easy for filing. The frames protect the transparency and are a convenient place to write lecture notes. Mount the basic acetate on the back of the frame. Using tape hinges, attach any overlays to the front. Avoid

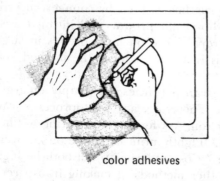
color adhesives

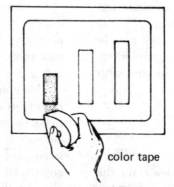
color tape

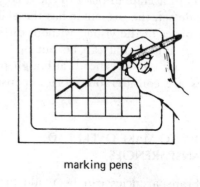
marking pens

Fig. 10-41. Using a razor blade or an X-acto knife, cut color adhesive sheets directly on the acetates. These tools, tapes, and color sheets are sold at art and office supply stores.

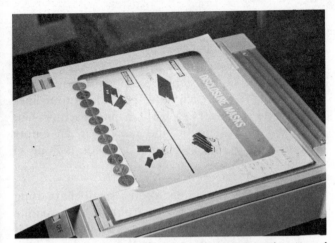

Fig. 10-42. Great discovery idea! With 11 pennies and a piece of stiff paper, you can make a mask for presenting overheads that won't fall off the projector when you pull it to the end of the list.

hinging from the top of the frame because that may conflict with the lens post on some projectors.

THE PRESENTATION

- Design a visual for each main idea. Limit each to fewer than a dozen words. Use no more than two pictures per visual.

- Look at the audience, not the visual.

- Give the audience a chance to look at and absorb each visual before talking about it. It's hard to read one thing and hear other verbal messages at the same time.

- Pace your visuals—don't go through them so fast they can't be comprehended, but don't leave them up so long that they begin to distract.

- Plan the places where you will initiate discussions. Shut off the projector at those intervals.

Wording
- brief
- key words
- incomplete sentences
- only support material

The caption's twenty-four words can be reduced to nine words for a visual

Fig. 10-43. Wording on visuals should be brief and concise. Don't use complete sentences. Key words will suffice, since the speaker will be filling in details.

SLIDES—AN OLD STANDBY

Perhaps the most effective visuals for agricultural and scientific topics are 35mm slides or "2 × 2's." They describe nature in complete and colorful de-

tail. Of course, tables, charts, graphs, maps, type, art, and other photos can be made into slides also. It's easy to make combinations, say, a chart which incorporates a photo or type with a map background. But the strength of 35mm slides lies in their ability to show the real people, places, and things the presenter wants to talk about.

Slides can be selected from a file beforehand and arranged in order to suit a specific audience. New slides can easily be added; old ones removed or updated. They're small, quite portable for traveling either with or without projector. If they're stored in plastic view pages in loose-leaf binders, you can easily find and retrieve them for future uses.

Fig. 10-44. A quick hint: After you have the slides for a talk all in order and oriented to go into the projector—mark across the top as shown. Now they will be easy to check at a glance; if any of them are out of position, the line will show it.

If the slide set you assemble will be used many times, it can easily be synchronized to run with a tape of the talk and even do the teaching without you! Why not do it with music at the opening, at interludes, and at the close? The audience will love it.

Slides are relatively inexpensive and easy to produce. Compared to all the other visuals, they're inexpensive to reproduce when you need duplicates. The expense is just in the film and development, once you have the camera and various lenses. With proper close-up photography equipment, the origi-

```
VISIBILITY STANDARDS
      FOR SLIDES
Up to 6 lines per slide,
    single spaced
Up to 25 characters per
   line, including spaces
```

Fig. 10-45. Visibility standards for slides.

nal can be as small as these typewritten words and as easy to prepare.

Art and type within an area 4 inches × 2¾ inches can be readable if it's done with a regular typewriter, provided you don't use too many words. While not the best quality type when enlarged on a screen, it's adequate. The cleaner, sharper, bolder, and bigger the type, the better. Therefore, for top-quality slides, you may still consider setting type larger than typewritten copy. Review the chapter on photography and experiment with your equipment to see how small the image area can be for flat art and type you prepare.

You can't rearrange slides as you talk and you can't write on them as you can with a chalkboard or overhead, but these are minor disadvantages. The major problem is the need for a darkened room. A little light hurts the visibility of slides. Darkness, on the other hand, makes notetaking impossible and sleeping natural. Keep the presentation lively and short to keep the audience awake.

There are many ways to produce graphic slides for titles and to support the information of your talk. Some of what you need to know is in Chapter 9. Remember, the artwork must be prepared on an area that fits the proportions of the actual slide, that is, 3:2. Therefore, 4- × 6-inch or 6- × 9-inch paper can serve as the art preparation sheet.

Here are a few simple methods you may use for preparing artwork for your slides. These methods vary in cost and quality. It's safe to say that the hour of care that you invest in preparing a quality slide will be more satisfying than a quick solution.

Phototype and filter. One of the more effective means of achieving bold, clear copy on a slide is through the use of phototype. Most university and newspaper printing facilities use phototypeset equipment and operators. With a minimum of direction on your part, material which you send in handprinted or typed will be returned to you camera ready. It will have proper margins, spacing, and no paste-up required. The typeface choices are numerous, and words needing emphasis may be set in a different face or a bolder version of the same typeface. As with any other slide, follow the standards for visuals—no more than 6 single-spaced lines per slide, nor more than 25 characters per line. The content of your visuals will be legible in practically any circumstance.

The copy you receive, black on white, may be converted to slides with color added by a glass or gelatin filter, or colored acetate between the art and the camera lens. Beware, some colors are dark enough to obscure the words and diminish legibility. Full-strength blues and reds are worse than yellows, greens, or oranges.

Mimeograph stencils. Type your message in the usual manner on the stencil. Keep within the 1⅜- × ⅞-inch area of a normal slide. This allows four lines of about 14 characters each on an elite machine. Cut the typed portion to fit, and place it in an empty slide mount. A cheap, usable word slide is done in minutes.

Type and tape. It's simple to do a small chart using paper and either the transparent or the opaque tapes that are available. In our example, the scales were typed on white paper. A rectangle of blue paper was glued in place, and the bars of tape were put on top.

Negative slides. Typewriter type and simple line drawings in black ink can be copied on black-and-white film. Single frames of the negative are mounted in slide frames. Color can be added, using a cotton swab and food coloring. For best results, use a high-contrast film, such as Kodalith. It eliminates dust shadows and edges of paper if you are using clip art with the type.

Superimposed titles. Using rub-on letters and photography, you can make title slides that rival the professionals. Select the best introductory slide you have, making sure to look for one with a dark, uncluttered area within the picture. This area could be a large shadow or the deep green of a lawn. You're going to have white lettering there. Buy a sheet of transfer lettering in a style suitable to your subject. Rub down the word(s) of your title on a sheet of white paper near the center.

The next step is the trickiest. You must estimate or measure how large to draw the "frame" of the image. Remember the proportions are always 3 units wide × 2 high. To help you estimate, look at the slide scene you'll use as a background and visualize the words in place. Draw the image frame so you'll know how far away the camera must be for the shot of the title. You'll focus slightly inside the lines you draw. Use Kodalith film for the shot. Get the Kodalith developed, then send the slide with the title and the slide with the background photo to the photoprocessor. Ask him/her to combine the images on another slide.

Sandwiched titles. You can make a slide similar to the previous example, but with black letters showing through the lightest areas of a photo. Do the lettering in the center of a white page. Position it with the same care. Photograph it with any slide film. After it

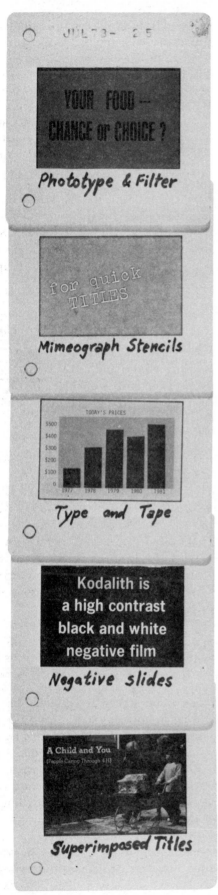

Fig. 10-46. Five methods of preparing artwork for slides.

Fig. 10-47. Superimposed titles to produce title slides add a touch of professionalism and quality to the homemade slide set. Here are five ways to do this.

has been processed, simply remove the title from its frame and slip it into the same frame as the background scene.

Colored paper. With scissors, rubber cement, and construction paper, you can create colorful titles. To make uniform letters easily, cut several rectangles the same size (those for "I's" will be narrower, and those for "M's" and "W's" slightly wider than the rectangles for other letters). If a letter is the same on both sides—"H," "O," or "T," for instance—fold the paper down the middle and make only half the usual number of cuts. Use contrasting colors for lettering and backgrounds.

Purchased lettering (3-D). You can buy many types of inexpensive paper, cardboard, and plastic reusable letters in stationery stores. Combined with cutout drawings or photographs and mounted on attractive backgrounds, these letters make slides look professionally produced.

Felt-tip markers. Used on newsprint pads or colored paper, felt-tip markers are excellent for illustrating slides. You can make original drawings or copy them from magazines and newspapers. You don't have to be an artist to use this technique. If you have difficulty keeping your hand steady, you may find it convenient to use templates or stencils to help you in lettering. With practice you can soon be a real pro.

Animation cell. By using the same machinery you use for overhead transparencies, you can make slides like those used for animated cartoons. For color, you will also need paint. For best results, use paints, such as Cel Vinyl, made for professional animators. Acrylic and gouache paints will also work, but not as well. All are available through your art store. Prepare the overhead artwork as you would normally, but allow generous margins. The reverse side of the line art that you put with the words can be painted. Start with the light colors and the smallest areas of the picture first.

If you allow them to dry, you can paint the larger areas of color easier. The painted acetate then can be placed on another background, such as a large photo, a scene of colored paper, or a distinctive texture (for example, wood, burlap). You'll avoid shadows if you place the assemblage under a sheet of glass when you're shooting the slide.

These are only a few of the many ways to make your own slide artwork. There are a variety of inexpensive lettering guides and adhesive-backed letters to help you prepare professional-looking titles. Many of the ideas presented in this section can also be used in making posters or simple displays.

Many other effects are possible, using professional help and equipment. If you've an idea of how a slide should look to be perfect for your message, ask a pro. Chances are there's a way to accomplish it. It may surprise you how simple and inexpensive it can be; then again, even if it would cost too much, it won't hurt to ask.

◀ A Rear Projection Screen You Can Make ▶

Not only is this rear projection screen easy to make, but construction requires no expensive or hard-to-get materials.

USE IT IN MANY WAYS

Since it's not necessary to darken the room completely when you use rear screen projection, your audience can take notes, and you can supplement the slides with flashcards, a newsprint pad, or other visual techniques.

A rear projection screen is versatile. Use it with group meetings in homes. Use it in an office to show slides to visitors. This is much easier than getting out a large screen and darkening the room. Use it in exhibits. Rear projection screens are excellent attention-getting story-tellers. Use rear projection exhibits at office open houses, in county fair booths, in window displays, or at any other place where you want to tell your story to the public.

The larger the screen area, the greater should be the distance between projector and rear screen.

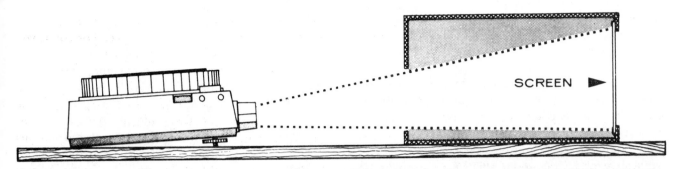

Fig. 10-48. Position the box both ways to know how wide to cut the front and back windows.

Check distance before cutting out screen area in the box. Vertical slides can't be mixed with horizontal ones unless your screen is square (which we don't recommend).

Load slides into the projector backwards for proper viewing on the rear screen. (This isn't necessary with a typical commercial rear screen unit, which contains a mirror).

Motion pictures won't work on this simple type of rear projection screen, since the image must be reversed by a mirror for proper projection.

Mark projector and screen positions on the table with bits of masking tape to simplify realignment if they are moved accidentally.

MATERIALS NEEDED

First, find an appropriate size corrugated cardboard carton. Then, get a marking pen or pencil for outlining the screen area, a metal-edged ruler, a razor blade, an X-acto knife or similar sharp-cutting instrument, and material for the screen.

A wide variety of materials is available for making the screen. If the screen is a single-use project for a special display, you can probably manage with just a piece of tracing paper. But to play it safe, tape the paper behind a sheet of glass or transparent acetate to prevent its tearing. For a more permanent rear projection screen, you might try using translucent acetate, obtainable in art and stationery stores. This comes in several types and thicknesses. Or, you could use Plexiglas with one frosted surface. In a pinch, you could use a translucent shower curtain liner from a department store.

HOW TO MAKE THE SCREEN

Select any size box that's appropriate for your needs. First, position the screen on the front of the

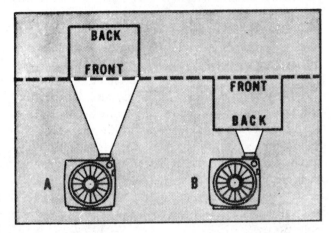

box. With the box placed upside down (flaps on bottom), project a horizontal slide onto the front of the box. Adjust the projector until the picture covers the proposed screen area when in focus.

Next, draw a rectangle inside the picture area and ⅛ inch *smaller* on every side. A marking pen makes an easy-to-follow guideline. Keeping the front on the same line as before, turn the box around so that the back of the box is now closest to the projector. Draw a rectangle around the projected picture, but this time make it ⅛ inch *larger* on every side than the actual image. Cut out the two rectangles.

For a neater finish to the large screen opening, you may want to frame the edges with masking tape. Rounding the corners slightly gives an interesting effect. You can spray paint the box, or you can use a textured, adhesive-backed decorator paper to achieve an attractive and professional look. Attach the screen material inside the box by gluing or taping it in place.

For a really permanent rear projection screen, use a wooden box and add a handle for easy transporting.

Many persons are effectively using homemade rear projection screens. Try the simple version described in this unit and make better use of your slides.

Maps are one graphic element everyone uses and relates to, because each person is somewhere, possibly wishing to be elsewhere. Because we think we know all about maps, this is a good opportunity to stretch our thinking about creative graphics. Almost everything suggested under this heading could be transferred to use in other graphic situations. First, however, what is essential to all maps?

They all represent imaginary lines? Yes. They're always read with north at the top? Almost always. They're all flat, having only two dimensions? Usually. They rely on symbols, lines, and words for specific meanings? True.

Now, how can things so mundane do what we want graphics to do? How can they (1) grab attention, (2) set concepts about the material, (3) enhance the written work, (4) clarify difficult points, (5) maintain reader interest, and (6) provide humorous relief? Try some of the following ideas and any that they inspire.

1. Use bold outlines.
2. Blacken all but the area you want to discuss.
3. Provide depth illusion with a shadow edge.
4. Pull up an area.
5. Drop a place.
6. Stack states, or separate them with a knife.
7. Personify a place.
8. Pattern a noteworthy place.
9. Make places move.
10. Make maps traditional or unconventional, as suits your subject.
11. Illustrate the features of an area.

A minimum of information still says "map." Size is no problem; in fact it's a tool to define more than physical relationships. Twist maps for drama or topple them for attention. And remember—if a map is of a state or other area with which the audience is already familiar, you don't need to waste space labeling the map.

Any flat object can be cut to the shape of a map. Then the meaning shifts to the inhabitants, their products, the scenic features, or whatever. Cut a photo of the North Woods to the shape of Minnesota, and it says forestry, or possibly tourism within the state. Take a photo of a happy crowd and cut out the shape of the United States. How about an angry crowd?

What if a map were cut out of cloth, corrugated lumber, or leather? Could a place be poured in concrete—formed in a puddle—polished on Plexiglas? Then what? Make a photo of it. Blow it up for an exhibit. Use it as a slide. Publish it. The point is, you have the power to make a map or any other graphics communicate.

As you can see, the graphic design principles define the limits between communication and unfettered creativity. Don't be afraid to use graphics; the benefits are too great to ignore.

Photo/Art Credits

All drawings were prepared by *Dale Reed* and *Henri Drews,* both at the University of Minnesota, except for those provided by:

Don Breneman, University of Minnesota: Fig. 10-19.

Dave Hansen, University of Minnesota: Figs. 10-12, 10-14, 10-15, 10-16, 10-20, 10-21, 10-29, 10-32, 10-33.

U.S. Department of Agriculture: Figs. 10-26 and 10-27.

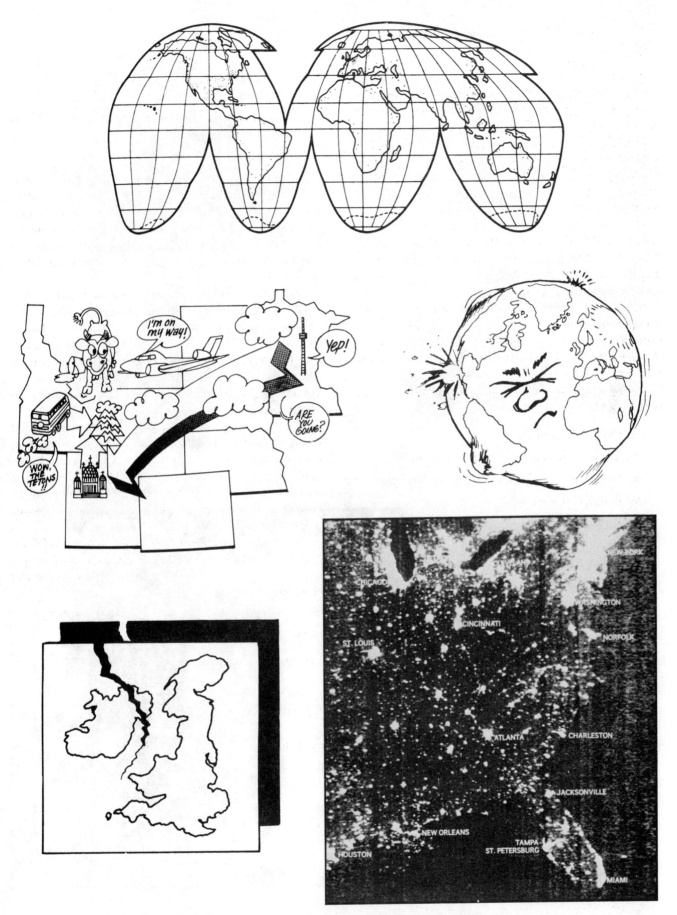

Fig. 10-49. Be creative and use maps in unconventional, as well as traditional, ways.

Fig. 10-49 (Continued)

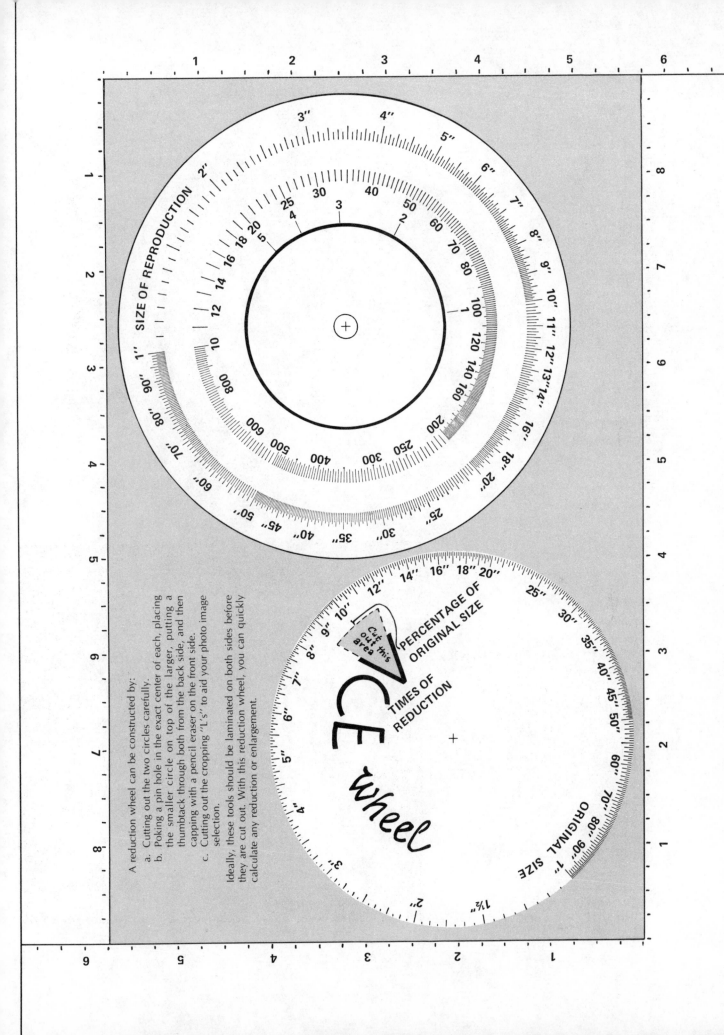

A reduction wheel can be constructed by:
a. Cutting out the two circles carefully.
b. Poking a pin hole in the exact center of each, placing the smaller circle on top of the larger, putting a thumbtack through both from the back side, and then capping with a pencil eraser on the front side.
c. Cutting out the cropping "L's" to aid your photo image selection.

Ideally, these tools should be laminated on both sides before they are cut out. With this reduction wheel, you can quickly calculate any reduction or enlargement.

SIZE OF REPRODUCTION

PERCENTAGE OF ORIGINAL SIZE

TIMES OF REDUCTION

CE wheel

ORIGINAL SIZE

Cut this out area

11 EXHIBIT AND POSTER DESIGN AND PRODUCTION

The exhibit is a unique form of communication that is growing in popularity. With the increasing cost of travel, many people are seeing the efficiency of having an exhibit communicate their message over and over again to large audiences.

An exhibit can include many media: film, television, photography, print, and others.

Unlike some media, which take the audience from A to B, to C, and on to a logical ending, an exhibit allows the viewers to start or to end anywhere, spending whatever time they wish with any particular element.

Most individuals give an exhibit a quick once-over at first and then look further only if it arouses their interest.

Exhibit viewers rarely devote as much time to an exhibit as the planners would like them to. They generally take from a few seconds to several minutes to look at an exhibit.

Competition for the viewers' time and visual attention is steadily growing. Today, the competition is bigger, brighter, and more colorful than ever. At one time, educational exhibits were confined to store windows, bank lobbies, and Grange halls. We currently find exhibits in increasing numbers in shopping malls, hotel and motel lobby areas, and larger community centers. Many county fairs now have spacious new facilities that accommodate hundreds of displays.

An exhibit must be well planned, be visually exciting, yet brief and to the point to compete effectively for the viewers' attention.

◀ Planning Your Exhibit ▶

PURPOSE

Many times the purpose of a communication project is decided before the medium is selected. The purpose may be to *teach* some unit of information, to *show relationships* between various elements of a process, or to *promote* an event or an idea.

An educational exhibit, if it's to be effective, must convey to the viewers something they didn't know before viewing the exhibit. You may wish to modify their behavior in some way, or influence their at-

titudes or beliefs. Decide if you want them to take a specific action, change opinions, or just be aware of some new information.

SUBJECT

Select a subject that will have personal appeal to a large portion of your audience. Timing is also essential. For instance, a vegetable variety exhibit would interest more people before spring planting time than it would in the fall.

Prepared by **Harry A. Carey, Jr.,** extension exhibits specialist, The Pennsylvania State University.

Choose a subject that is specific. A broad subject will be harder for you to cover adequately and tougher for your viewers to comprehend. A simple subject is easier to design, particularly if you limit the number of main points to three or four.

Be sure to include only material that provides a real contribution toward your purpose. Look critically at the information as your viewers will see it. Hit them hard with the main points of your subject and eliminate the details. Detailed information can be printed in a handout, or you may provide an address or a telephone number where it can be obtained.

AUDIENCE

Decide exactly who your audience will be. Many of the decisions involving design will be affected by the age, background, educational level, and lifestyle of your audience. Young people generally enjoy bright colors in unusual combinations and wild lettering styles, while these may be offensive to older, more conservative individuals. It's easy to find several people who align with a specified audience, and you can ask them for opinions on your design ideas.

Consider the location(s) where the exhibit will be displayed. What kinds of people will be there? What are their interests? How familiar will they be with the information covered in your exhibit? What can your exhibit do for them?

Exhibits are intended to be different from newspapers, bulletins, and magazines. Messages should be conveyed in a *visual* manner (see Fig. 11-1). The more printed copy included in an exhibit, the less time viewers will spend on it. Keep the copy to a minimum and the visuals to a maximum. Be sure that the visuals you select really contribute to the communication of your message.

Live or real objects. Because of the convenience of obtaining, mounting, and displaying other visuals, we often overlook using real things. Live animals always attract attention, and plants or other real objects are usually superior visuals if we can use them.

Models. Whenever the size or visual limitations of a live or real object are prohibitive, a model may be helpful. A model can provide a miniaturization of a much larger object, or an enlarged version of a smaller object. A model can allow the viewers to focus on the important parts by eliminating unnecessary details or visual obstructions.

Photographs. Photo enlargements provide a realistic look at an object or a situation. The photos should eliminate distractions and irrelevant details and zero in on the areas that support the message to be communicated. This medium can condense large objects or enlarge small ones. Avoid having many small photos . . . use fewer but larger prints.

Projected images. Motion pictures, slides, filmstrips, overhead transparencies, and television can be useful as exhibit visuals. Common sense must keep the designer from expecting the audience to spend much time viewing this type of visual. Just because the pictures are moving or changing, or they are accompanied by sound, doesn't mean the viewers will spend much more time watching. Unless they have a strong interest in your subject, you can expect them to give your exhibit only two or three minutes, maximum.

An important consideration relating to projected images is the image size. While it may be desirable to have a large image, you'll find that the larger the image size, the less brilliant it becomes. Since exhibits are usually displayed in well-lighted areas, it's normal for the image on the screen to look weak unless it's kept fairly small (no larger than 18 inches), or unless you provide a canopy to eliminate the ambient light.

Fig. 11-1. Tell the story with visuals. The exhibit on the right has too much copy, not enough visuals.

Illustrations. This medium differs from all the preceding in that the creator can completely eliminate unwanted detail, exaggerate portions to provide emphasis, and communicate a message in its most basic form. While the appearance should be attractive, it's not necessary (it's often undesirable) for the illustration to appear as a beautiful work of art.

Graphics. Charts and graphs can be helpful visuals in showing changes, relationships, and differences. Be sure to choose the right type of graph for your message, and keep it simple in design.

Other graphic visuals include large or unusual letters, words, shapes, and design patterns that assist in attracting attention and communicating a message (see Fig. 11-2).

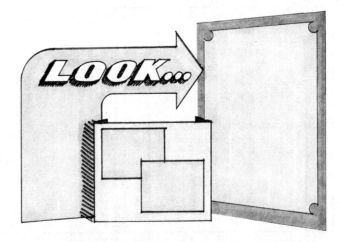

Fig. 11-2. In addition to the usual visuals, large or unusual letters, words, shapes, and design patterns can assist in attracting attention and communicating a message.

◀ Exhibit Design and Construction ▶

Exhibit design is an attempt to organize the basic shapes formed by the visuals, the copy, and the unused portion of the background (negative space) into an interesting and balanced unit. In such an arrangement, the elements themselves are subordinate to an overall unifying plan. This plan should be pleasing to the eye, yet simple in concept, and it should facilitate the communication of the message.

An exhibit designer must remember that while aesthetics are important, the art form or level of beauty is important only as a basis for communicating and clarifying the message.

The designer should have some knowledge about where the exhibit will be located. The theme of the event, the style of the exhibits, and the nature of the exhibit location affect design. The designer should

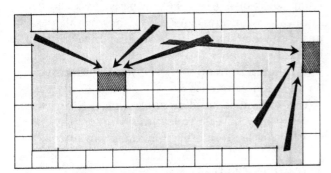

Fig. 11-3. Consider traffic flow and viewing distance. With large lettering and visuals, you can increase legibility and attract attention at a greater distance.

consider traffic flow, viewing distance, and whether or not the exhibit will need people working it (see Fig. 11-3).

Step 1

The exhibit designing process begins by gathering relevant facts and conditions or limitations about the exhibit: its purpose, the subject, the audience, its location, and the resources you have to work with. Write them down. You usually begin with much more subject matter information than a simple direct design will allow, and then organize it into some logical order that the viewers can follow easily. This information is boiled down into simple phrases or brief statements. Don't hesitate to delete information of lesser importance.

Make several attempts at a short, catchy title that identifies the exhibit and gets viewers involved. Use the verb in the active form, and be sure the audience can relate to and be invited by the title.

Step 2

In looking at this orderly list of statements, you must then consider the opportunities to support, clarify, or explain them with visuals. Jot these ideas down on paper. Consider a variety of approaches for each visual, but aim for visual ideas that are simple and to the point (see Fig. 11-4).

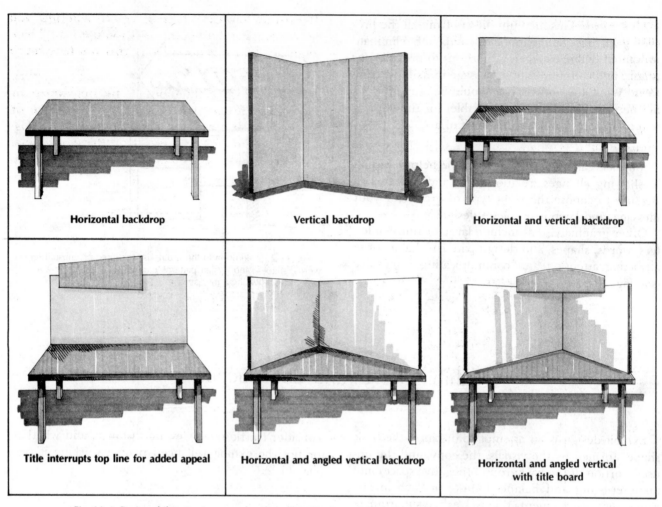

Horizontal backdrop	**Vertical backdrop**	**Horizontal and vertical backdrop**
Title interrupts top line for added appeal	**Horizontal and angled vertical backdrop**	**Horizontal and angled vertical with title board**

Fig. 11-4. Basic exhibit structures can be altered to complement the message and to increase the viewers' attention.

Say, for instance, that your exhibit would benefit from a visual that relates to horses. List a number of alternatives: a group of horses, one horse, half a horse, a horse's head, a horse's hoof, a horseshoe, a bridle, a saddle, a riding boot; the list can go on.

Now put yourself in the place of your viewers. Which visual alternative is most appealing? Which best communicates your message? Which is the least complicated?

Step 3

At this point you can begin to relate the visual elements to the overall design. Sketch out some overall design ideas. Asymmetrical, or informal, balance is nearly always more interesting than symmetrical, or formal, balance. The visuals and copy "elements" should be grouped, rather than distributed evenly over the background. This will allow larger areas of negative space, and the exhibit will look less crowded. Arrange the elements so the exhibit will read from left to right and top to bottom.

If you reach a point where ideas aren't flowing readily, take a break and come back to the project at a later time.

Step 4

After allowing your initial ideas to digest, take a fresh look at the title and those brief statements. Be sure they contain only the information that makes a real contribution toward your purpose. Rethink the visual selections. Are they the best choices to communicate your message? Does the overall design unite the elements into a simple, pleasing communicative unit? Does the design have a "shock factor" to attract attention? (See the following section on attention-getters.)

You may need to allow more time for digestion, and you may have to repeat this step, or parts of it, several times before everything falls into place.

ATTENTION-GETTERS

A good design can unite the various elements in

such a way that the overall appearance is attractive, inviting, and worthy of attention. If the viewers don't bother to look at it, they won't get your message. A good choice of visuals can be attention-getting, but coupled with one or more "attractants," your exhibit can draw considerable attention.

Size. Keep your visuals and lettering large. If your entire exhibit is small, then you must limit your selections in order to keep them as large as possible.

If surrounding exhibits have ½-inch lettering, you have a considerable advantage if you use 1- or 2-inch lettering. This may also require that you have much less copy (see Fig. 11-5).

Shape. Many exhibits appear to use only square or rectangular shapes. The use of a round, oval, or other unusual shape can draw attention (see Fig. 11-6). Backgrounds need not always have horizontal and vertical lines, even though they are convenient.

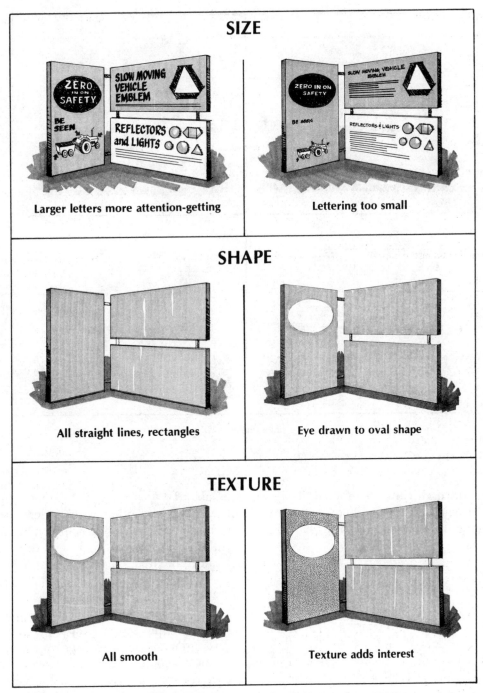

Fig. 11-5. Visuals alone are attention-getters; but certain size, shape, and texture features can increase the attention-getting capacity of an exhibit.

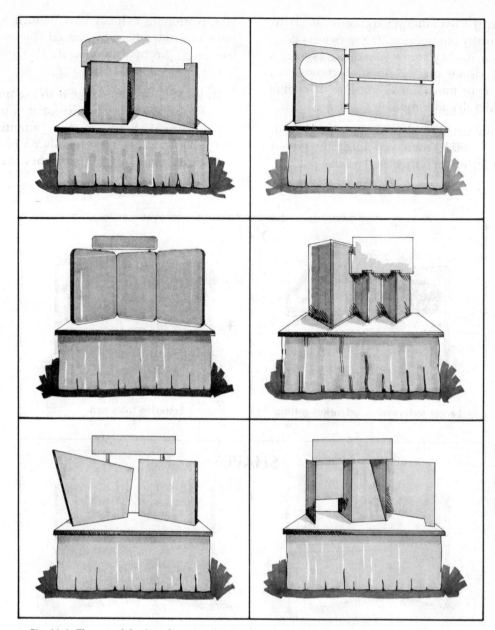

Fig. 11-6. The use of depth and varying shapes helps draw viewers to the exhibit for further study.

Try to use the third dimension in your design. Two-dimensional objects, such as photo prints, can be more interesting if they are spaced away from the background surface.

Texture. If most parts of your exhibit have smooth surfaces, it may be advantageous to add a rough or textured area. This is easily accomplished with rough wood, corrugated paper, burlap fabric, and the like. You can even give this effect by partially painting an area with aerosol spray or a paint roller, or by striping with a brush or felt-tip marker.

Color. It generally is wise to be cautious with the use of color, particularly on the background of an exhibit. Too bright a background can detract from the visuals and other exhibit elements. Usually two or three colors are adequate. Brighter colors are useful in attracting attention or in drawing the viewers to a certain area.

Motion. Motors of many varieties can be used to achieve animation. These can be purchased at an electric supply center. Department stores often discard animator motors along with exhibits when they are outdated. They are often pleased to donate them to a worthy cause. Live animals also provide motion, along with getting considerable attention, when they are part of an exhibit.

Light. Make sure your exhibit is well lighted. It usually takes about 150 watts per 4- × 4-foot area. While floodlights are used to light broad exhibit areas, intense spotlights can draw attention to specific exhibit areas.

Moving, flashing, and blinking lights can attract attention, but common sense must dictate their intensity. Incandescent lighting intensity can be reduced by using lower wattage bulbs or a rheostat.

EXHIBIT DESIGN TIPS

Review the following suggestions any time you are working your way through the exhibit design process.

1. *Be sure the design is basically simple,* with few elements and much negative space. The most common error of exhibit design is too many elements or too much copy that gives the viewers the impression of being cluttered and difficult to comprehend (see Fig. 11-7).

Fig. 11-7. The design should be basically simple, with few elements and much negative space. The top exhibit is too crowded.

2. *Make certain that the exhibit "reads" well*—left to right, top to bottom. The title, the visuals, the copy, and the overall design should have an obvious message, with all the elements contributing to its communication. Does the exhibit accomplish the purpose that was intended?

3. *Put yourself in the position of one viewer* from your specified audience. Imagine the exhibit in its intended location. Is there a strong attention-getter? Does the message come through loud and clear? Is that message positive and worth your time?

EXHIBIT BACKGROUNDS

A wide variety of materials is available for use as exhibit backdrops. Before selecting one, you should answer some basic questions concerning the exhibit. These include:

● What will be the final size of the backdrop?

● How small must it be packaged for transport and storage?

● How heavy will it be?

● How sturdy and durable must it be?

● How easily can visuals be mounted on it?

● What color should it be?

● What are the resources necessary, such as initial cost and time preparation?

Here are some commonly used materials:

1. Plywood, available in ¼-, ⅜-, and ½-inch thick, 4- × 8-foot sheets—heavy in weight.

2. Masonite, ⅛- and ¼-inch thick, 4- × 8-foot sheets—heavy in weight.

3. Plywood paneling, ³/₁₆-inch thick, 4- × 8-foot sheets—moderate in weight and sturdy.

4. Styrene and urethane foam, ½-, 1-, and 2-inch thick, 4- × 8-foot sheets—very light.

5. Foam board, ¼- and ½-inch thick, 4- × 8-foot sheets—very light and good as a surface.

6. Corrugated cardboard, various sizes—very light, but surface must be coated.

7. Posterboard, 14 and 28 ply, 28 × 44 inch and 30 × 40 inch—light and good as a surface.

Other wood, paper, plastic, and cloth materials are available. The preceding list includes those that are most widely used.

Limited storage space and smaller auto sizes increase the necessity for designing exhibits that are lightweight and can be squeezed into smaller packages. Foam board is a lightweight product that can be folded to pack into corrugated cartons or envelopes for easy transport and storage.

Lightweight paper and foam products can be easily used as backgrounds for tabletop displays. They need additional support or framing when used on a larger scale.

The large, free-standing exhibit backgrounds may be constructed with any of the aforementioned materials, provided they are adequately supported. Plywood is sturdy when framed with 1- × 3-inch pine lumber, and sections can be coupled together with bolts through the framing. The reverse side (non-grooved) of plywood paneling makes a good surface. The grain can be filled, or simply sanded and painted with several coats.

LETTERING

The lettering brush makes fine letters if you've developed the skill or can hire a professional, but often the available resources are prohibitive (costs or skill availability). Tempera poster paints are easy to use, and although they dry quickly, they are not waterproof. There are poster paints available that dry quickly and will withstand some weathering. Sign enamels are weatherproof and durable, but they have a glossy finish and need 24 hours to dry.

Speedball pens are more easily mastered than the lettering brush. These inexpensive pens work with many kinds of inks.

Art supply stores offer a wide assortment of stick-down and rub-on, or transfer, letters. The placement and spacing of these letters still requires a certain amount of skill and time.

Exhibit titles should have letters at least 3 inches in height, and a line thickness roughly one-sixth the height of the letters. Larger sizes are usually preferable. The lettering size in the copy of an exhibit should be no smaller than 1 inch.

Lettering visibility is affected by:

1. *Size.* Use lettering that is as large as possible.
2. *Line thickness.* The lines that form the letters should approximate one-sixth the letter height.
3. *Style.* Choose a style that is bold, easily read, and not too fancy.
4. *Contrast.* The letters shouldn't blend into the background. Use light-colored letters on a dark background and dark-colored letters on a light background. Don't force your viewers to get out their bifocals.

WORKING AN EXHIBIT

Many public exhibition areas become cluttered with disposable food and drink containers, handout literature, etc. Be sure that someone has the responsibility for keeping your exhibit working and the immediate area neat and clean (see Fig. 11-9).

Exhibits need attention in other ways also. Lightweight exhibits can be bumped out of alignment, and display materials that are handled often need to be repositioned. Handout leaflets should be kept in neat stacks, and the supply replenished as needed.

Individuals assigned to work an exhibit should be

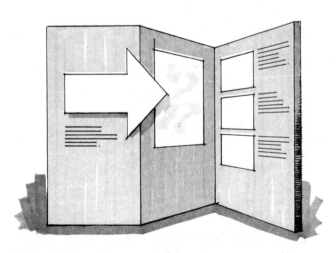

Fig. 11-8. Because most structured panels are manufactured in rectangular sheets, it's common to see exhibits with horizontal and vertical lines. Any effort to break up this horizontal and vertical effect will pay rich dividends in terms of increased audience interest.

prepared to answer questions concerning the exhibit subject and to provide information about the sponsoring organization. They should also be able to give food, drink, and restroom information. It's often appropriate to have reference materials available to assist them in answering the harder or more technical questions. A pencil and pad should be kept handy so that requests that can't be filled immediately can be recorded.

Persons who work an exhibit should be neat and well-groomed, appropriately dressed, and properly identified with a name tag or other means. They should be interested in assisting the viewers by greeting them, helping them make new friends, and answering their questions. This responsibility is demanding, requiring many hours of standing, while looking enthusiastic and fresh. Plan on plenty of help, so no one will be required to work beyond his/her capacity to the point of looking tired and unfriendly.

Fig. 11-9. There are various reasons for having individuals work an exhibit. Keeping the exhibit working and the surrounding area neat and clean are the minimum requirements.

SUMMARY

- Remember that onlookers usually view an exhibit for only a short time—from several seconds to several minutes.

- Because of the increasing visual competition, you must have an attention-getter and a quickly understood message in your exhibit.

- Decide first on your purpose, subject, and audience; then design the exhibit with those things in mind.

- Do a good editing job, allowing only the most relevant points as part of the message.

- Tell your story basically with visuals.

- Leave plenty of space in your exhibit; it shouldn't have that discouraging, crowded look.

- Select appropriate materials. It may be necessary to pack your exhibit in a convenient, transportable package, which must be reasonably lightweight.

- Keep the lettering to a minimum in quantity, but plenty large in size.

- Have the display set up on time. Keep it looking good throughout the showing.

- Make sure persons working the exhibit are neat and well-groomed, enthusiastic, and above all, interested and helpful.

◀ Judging Educational Exhibits ▶

The most important reason for preparing an exhibit is to "tell a story" for educational, publicity, or promotional purposes. A number of factors must be successfully executed if an exhibit is to convey a message in a meaningful way to a potential audience. The score sheet in Fig. 11-10 lists some of these factors. Becoming more widely used each year, this score sheet differs from others in that it emphasizes attracting attention (20 points) and conveying a message (30 points). These areas cover the prime reason for preparing an exhibit in the first place. A few comments on the use of this score sheet may be helpful.

1. Make a score sheet for each exhibit to be judged.

2. Examine all the exhibits and give each a score for "Attracts Attention." Spread your ratings throughout the 20-point range. If some exhibits are weak on this point, don't hesitate to

mark them down. They may rate high on other points.

3. Rate all exhibits on "Arouses Interest"; then on "Conveys Message."

4. Add the scores on each score sheet and review the exhibits in order of placement.

5. If ties occur in top, crucial placings, perhaps review the tied exhibits and adjust the scores.

6. Judging can be an educational experience for the exhibitor, particularly if you add a few notes on the strong and weak points as you judge them.

SCORE SHEET FOR JUDGING EXHIBITS

	Points	Score
ATTRACTS ATTENTION	20	

Makes use of size, shape, texture, color, motion, and light. While attention-getting is important, the attention should be favorable.

| **AROUSES INTEREST** | 10 | |

Encourages additional study. Gives a personal appeal to the type of audience for whom the exhibit was designed.

| **CONVEYS MESSAGE** . . . . | 30 | |

The message should be conveyed quickly—in about 30 seconds. (However, additional detail of importance should not be discounted.)

Viewers should leave the exhibit knowing something they did not know before they came. The message should be understandable to the intended viewers.

| **DESIGN** . . . . | 20 | |

Elements of the exhibit should be pleasingly placed to give a sense of unity to the whole. The design should enhance the organization and the readability of the topic. The message should be part of the design.

Simplicity, the grouping of elements, open space, and a lack of clutter, or crowding, are important.

| **ORIGINALITY** . . . . | 10 | |

Shows evidence of creativity.

| **QUALITY OF DISPLAY WORK** . . . . | 10 | |

Is neat. Is well constructed for the purpose. (This does not imply that expensive materials must be used.)

| **TOTAL** . . . . | 100 | |

Fig. 11-10. This score sheet highlights the relevant points of a good exhibit.

◀ Some Poster Design Ideas ▶

Posters are commonly used to advertise an event, such as a meeting, a show, a contest, a fair, a dance, a picnic, a flea market, or a public sale. Posters are also used to promote a political candidate, a product, or an educational idea.

Posters differ from flipcharts, flash cards, and other teaching aids in that they have no one explaining them or calling attention to them. A poster must stand alone. No one will call attention to it, identify its purpose, or stimulate onlookers into further action.

The average person who glances at a poster usually looks only long enough to identify it. If it's graphically exciting, or if it deals with a specific interest of the viewer, chances are he/she will look at all the details and get the complete message.

Posters are normally produced on paper or posterboard, but they can take other forms such as promotions printed on trucks, billboards, T-shirts, and bumper stickers. The form your poster takes should depend upon the audience you wish to reach.

WHAT AND WHO

Begin by deciding specifically what it is you want to say and exactly who your audience is. Put these ideas down on paper. Continue to search for better word and image choices in an effort to express your meaning more clearly, simply, and precisely.

Consider your message from the audience's point of view. What is the age, educational level, background of this group of people? Is the choice of words and images appropriate for them? Will these words and images catch their interest? An understanding of the audience will also be of value in making design decisions regarding the color, type style, illustration, and basic layout.

Jot down words and phrases and make rough sketches that describe or clarify your message. Begin to arrange these in a logical pattern. Select the most important words; pencil them into a rough sketch in large letters. Rethink what you've just done. Did you really select the key words from the onlookers' point of view? What's important to them?

THE ILLUSTRATION

Start to develop the illustration. By now you should have several rough ideas to work from. Develop an idea that will appeal to your selected audience. Explore various approaches to the illustration, but aim for a visual idea with impact, simplicity, and conciseness. The illustration must reinforce the message, or better yet, carry the message.

If the message that you've selected for your poster relates to vegetable gardening, you can list a number of illustrative approaches: an entire garden, one row of plants, a group of plants, one plant, an array of vegetables, one vegetable, a slice of vegetable, a jar of canned vegetables, a hoe or a cultivator, a person holding a vegetable, an expression of satisfaction on a person's face; the list can go on.

Now put yourself in the place of your intended viewers. Which idea best communicates the message? Which idea is most appealing? Which of these ideas can you execute without risking confusion? Which of these illustrative ideas has greater potential as a design element in adding impact and drawing the attention of the audience? Seek a simple solution to these questions.

A basic poster design often makes use of a single illustration that plays an important role in getting individuals to notice the poster, as well as in communicating the message. Therefore, the final choice and form of the illustration is critical to the effectiveness of the poster.

LAYOUT AND DESIGN

One effective approach to poster layout and design is to use a large, brief title, along with an illustration, as an attention-getter. Other necessary details may then appear in smaller type. This approach assumes that the large title and illustration grabs the viewers' attention and lures them in to study the details more closely.

Good poster design organizes the basic shapes formed by the title, the illustrations, and the body copy, along with the open space, into an interesting and balanced unit. This basic arrangement should enhance the communication of the message, be simple in design, and be pleasing to the eye.

DESIGN GUIDELINES

Keep the *message* foremost in your mind throughout the project. One to five key words should appear in large type and be legible at a distance. These words should express the core of your message and have impact upon the audience.

The major *design elements* are the blocks of space occupied by the title, illustration, and the body copy. It's easier to plan a successful design with three elements than it is with many elements. One of the elements obviously should be larger than the others. This contrast in size adds to the interest created by the design.

Poster design elements are more effective when *grouped* than when spaced over the entire poster area. Therefore, you should try to avoid creating multiple elements. For instance, a three-word title might be printed in a design with the three words tightly fitted together so they would appear as a single element. Another treatment would be to space the three words apart so they appear as three separate elements. These three elements, along with an illustration and a block of body copy, would increase the total number of elements to five.

Leave plenty of *space* between and around most elements, with extra space along the edges of the poster. Avoid the look of crowding.

Allow for several fairly *large areas* of open or unused space. A design begins to look crowded whenever the open space areas fall below 20 percent of the total area. Many successful posters have 30 to 40 percent open space.

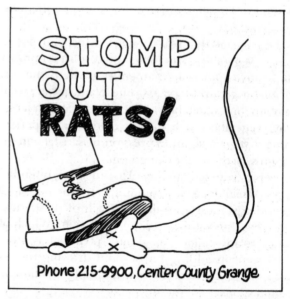

Fig. 11-11. A selection of poster ideas. Note that the three on the right give a variety of treatments to the same topic.

The *configuration* of the open space is just as important to the impact of the basic design as the shapes formed by the lettering and illustrations. It's usually desirable to have various sizes and shapes of open space.

Balance is an important part of poster design. This refers to the relative "weight" of the visual elements. While it's usually undesirable to scatter the elements evenly over the entire poster area, it's usually ineffective, also, to crowd all the elements into one end or one corner of the area.

Balance is necessary to design, but it's often more effective if achieved by means other than by centering the elements. Asymmetrical, or *informal,* balance is usually more interesting, more fun to work with, more challenging. Informal balance is less static and monotonous; it suggests movement.

In a superior design, either the illustration or the lettering *dominates,* rather than an equal division between the two. A design with the illustration occupying more area than the lettering grabs a lot of attention, but you must take care to assure the illustration is supportive of the message. People often resent trickery, such as using an attention-getting illustration or words that are misleading and not supportive of the communication of the message.

Horizontal lettering is normal, while vertical lettering, which is difficult to read, is not recommended. Therefore, an abundance of poster elements appear in the horizontal position. A large, *vertical* line or illustration can attract the viewers' eye and become a pleasing design element. This is also true of angular lines and shapes that contrast to the rectangular edges of the usual poster format.

Lines create flow and *direction,* causing eye movement. The juncture, real or imagined, between two or more lines can easily become a focal point, or center of interest, within a design. This means that as you become adept with the use of lines and direction, you have increasing control over the eye movements of the viewers.

It's often desirable to develop a strong *center of interest* on the words or the illustration that is the quickly grasped key to your message. This reinforces the idea that communicating the message is the basic task that you set out to accomplish in the first place.

Avoid placing the center of interest in the geometric center of the poster area. Also, refrain from placing it too close to the edge of the design, or tightly crammed in a corner. Roughly, a *third* of the distance up, down, or in from the edge is much more desirable.

Designing is easier if the *body copy* is tightly edited and is presented in the form of a relatively tight package or block. Consider with care the information presented here. On the one hand you can't overlook certain necessary facts, such as time, place, etc., but you must scrutinize other explanatory information very carefully. If the amount of body copy becomes too great, it must be printed in very small lettering, or it will take up too much space and cause the poster to look cluttered and crowded.

Be certain that the choice of *colors* doesn't detract from the message and is acceptable to your selected audience. A related color plan is a conservative, safe approach, which seldom jeopardizes a design. A related color scheme uses colors that lie next to each other on a color wheel. Certain designs with specific audiences might benefit by the addition of a brighter, contrasting color. Older audiences often prefer the more conservative color plan, while younger audiences like wilder, contrasting, and even vibrating colors.

Thus, there's no one absolute formula that will guarantee you success, although many of the forementioned rules have proved effective over time. While these rules are useful and important, one or two of them can be broken, for good reason, and an effective design can still be accomplished.

LEGIBILITY

Many of the design decisions should be made with the overriding goal of legibility in mind. If words are difficult to read or an illustration complicated and confusing, the message won't come through loud and clear, even if the audience bothers to read it.

A number of factors influence legibility. These include:

1. *Size.* Letters and illustrations must be large enough to be easily recognized. A large poster, 22 inches × 28 inches, should have the title printed in letters that are 2 inches to 4 inches in height. A smaller format, say 14 inches × 18 inches, should have title letters 1½ inches to 3 inches high. Body copy is often printed in letters that are one-fourth or one-third the height of the title.

2. *Contrast.* This refers to the relative lightness or darkness of the elements and the background they are on. Maximum contrast is black on white or white on black. It is tiring on the eyes and shouldn't be overused. Minimum contrast, such as pale yellow letters on a white background, doesn't read well either. A combination just short of maximum contrast is desirable. This might include black or brown letters on yellow paper, white or pale yellow letters on

medium to dark (but not black) tones, or red letters on a beige or sand-colored background.

3. *Style.* This pertains to the character or tone of the graphics. Some lettering styles, such as Helvetica and Franklin Gothic, are very easily read, while others are too ornate or too fancy to be easily read. An illustration, also, can be too detailed to the point of confusion. Try to select styles that are simple, legible, and handsome, thus enhancing the message.

4. *Line thickness.* This must be reasonably bold for good legibility. Ideal line thickness for letters is roughly one-sixth the height of the letter. A 3-inch-high letter, with a line thickness of $1/16$ inch, isn't very visible. A 3-inch-high letter with a line thickness of ½ inch (one-sixth the letter height) is legible at 80 or 90 feet. Fairly bold line thickness is also necessary for illustrations to be readily visible.

MATERIALS

Poster messages appear on a wide variety of surfaces. If you intend to ink or paint the lettering or illustration, you should select a paper or cardstock that has a smooth, hard finish. A hard finish will be less absorbent and will allow you to produce sharp, crisp lines.

Heavy printing paper, cover stock paper, and poster paper are sometimes used for posters. Cardboard that's suitable for poster production is available under a variety of names. Bristol board, railroad board, showcard, illustration board, and posterboard are commonly used. They vary in thickness, in overall size, in the hardness of finish, and in price. Compare the available products and examine them closely. Select the material that best suits your project and your budget.

The lettering and illustrations may be produced on the poster with ink or paint, or they may be glued onto the poster surface. Felt-tip markers may be used. Be sure to select those with permanent ink. Other inks, such as India ink, may be applied with a metal-tip pen or a brush. Different kinds of paints are available, but tempera poster paints are the easiest to use, for they clean up with water.

Letters that are commercially available include rub-on transfer letters and pressure-sensitive paper, or vinyl, letters. Although expensive and time-consuming to apply, they're very precise, clean letters. Stencil guides may be used to outline the letter shapes on your poster. Be sure to fill in the letters where the centers are attached to the rest of the stencil.

SHOWING

You've done your best to achieve a good poster. Take care in its display. Even an excellent poster can be ineffective if displayed under poor conditions. Think of the places that your viewers might frequent. Where might they congregate or pass idle time?

Attempt to place the poster where it receives good visibility and isn't lost in a confusion of competitive clutter. Attach the poster securely, so it can't slide, fall, or wrinkle. It should be at eye level or above, and on a pleasant-looking bulletin board, wall, or some other structure.

The poster should have adequate lighting to be seen. Because fluorescent lighting will influence the color scheme, the poster may appear much different than when viewed under incandescent lighting.

One to three weeks is usually adequate lead time for posters to advertise an event. You may wish to check on their condition several times throughout this period to insure that they remain in top condition and maintain an acceptable appearance.

SUMMARY

People glance at posters and look further only if a quickly grasped message strikes their interest. You must carefully select the best key words and/or illustration to grab their attention.

The basic design should be simple, with few elements and much room at the margins, with some larger areas of negative space. The most common mistake in poster design is too much material scattered over the entire poster area, much like a newspaper. This gives the viewer the impression of being cluttered and difficult to comprehend.

Always keep the message and the intended audience in mind. You must communicate in order to succeed.

Photo / Art Credits

Harry A. Carey, Jr., The Pennsylvania State University: Fig. 11-11.
Peter A. Kauffman, The Pennsylvania State University: Figs. 11-1, 11-2, 11-3, 11-4, 11-5, 11-6, 11-7, 11-8, 11-9, 11-10.

12 MEETINGS THAT SUCCEED

◀ Planning Meetings ▶

Every successful meeting has a clear objective. No one attending should have the slightest doubt about the meeting's purpose. You'll need to ask how far the meeting is supposed to move the audience and in what direction. What specific action, if any, is the audience supposed to take following the meeting?

Clearly specify the target audience. How do you invite these persons? Will the invitation prepare them for what will happen at the meeting? Is there anything special about the members of this particular audience? Are they sponsors of the ideas or information to be presented? Are they a unique group in terms of income, farming specialty, nationality, status, or experience with the subject?

Is the topic timely for this audience? Is it what the audience will expect? Do individual items on the program fit together?

Specific answers to these questions will give you clues on *how to plan the message* for the program. You'll need to consider audience attitudes; whether the message will get and hold attention; and whether it will stimulate thinking, have credibility, provide or suggest a logical basis for considering or adopting facts or procedures presented at the meeting, and arouse interest toward eventual action.

PLAN FOR AUDIO-VISUAL AIDS

Many speakers use audio-visual materials to stimulate audience interest and response. As meeting arranger you'll need to provide the speaker with chalkboards, charts, posters, slides, flannelboards,

Fig. 12-1. Plan for audio-visual aids to stimulate audience interest and response.

models, and any other devices he/she wants to use. You may want to suggest some method for reaching the particular audience involved. Will your meeting place provide all the facilities the speaker needs?

For example, if the speaker plans "buzz" sessions, it's obviously important that the seats be movable. Or, will the speaker give a demonstration? Have you made arrangements to avoid interruptions? Can the speaker easily conduct a work session? How does the audience-stage relationship affect the meeting formality? Will you need to consider possible audience participation? Will the physical arrangements help keep the meeting moving? And don't forget to make a last-day check on facilities, equipment, and materials.

Prepared by **Hal R. Taylor,** former director of public affairs, U.S. Department of Agriculture.

REMEMBER THE AUDIENCE

The key to an effective meeting lies in the audience. First, get these people involved in planning the meeting.

When you get local people involved in the program, you are doing things with them rather than for them. It takes away some of the temptation to think in terms of what you think people should have, rather than what they want. Asking people what they want involves them in the planning itself and gives them a stake in the meeting's success. They're likely to work harder and to assume more responsibility for the aims of the meeting.

At or during the meeting, you can encourage favorable audience attitudes by looking after physical comforts. For example:

1. Is the audience comfortable? If chairs are so hard that the audience can't relax, it's a good idea to provide frequent stand-up breaks.

2. Did the meeting start on time? Can every member of the audience hear what's being said and see what's being done? Is there adequate lighting so people can take notes? Is fresh air moving into the room?

3. Are you able to alter arrangements quickly if necessary? It's sometimes impossible to anticipate all emergencies, and you may still need to depend on your skill and resourcefulness. Be ready to "play it by ear" and to improvise quickly if necessary.

EVALUATE CONSTANTLY

Although you may plan to find out after the meeting what the audience learned, you can also tell how the meeting is going while it's in progress. Watch how well the message gets audience attention and holds it. See if the program arouses interest and meets the objectives. Decide whether the message and supporting materials are too abstract or complicated, or whether the words, ideas, symbols, and thoughts are familiar to the audience. In other words, determine whether you're pitching your message to the audience. Look and listen critically.

If you feel that the participants left the meeting with unanswered questions, be sure to follow up with the information they want. It's always a good plan to follow any meeting with the subject matter information in releases to the public media or in letters to individuals or groups.

USE EFFECTIVE GROUP TECHNIQUES

In general, reserve group techniques such as buzz groups, problem-solving conferences, panels, role playing, and case studies to small groups. Large groups require special techniques, which include controlled participation, problem census, pre-questions, listening teams, information conferences, and subgroup work projects.

Of course, selecting the best technique for the size of the group is often a puzzle. The solution frequently depends on the participants' attitudes and abilities and the leader's skill.

◄ So You're a Conference Leader ►

A conference leader's self-image affects his/her attitude toward the task and can also affect his/her success in working with members of the group. Success or failure of a conference doesn't depend upon a leader's knowledge of the subject. In fact, the leader need not know more than the group about the subject. The effective conference leader will create the proper setting for learning.

THE EXTREME TYPES

One extreme type of conference leader considers himself/herself an expert on the subject under discussion. Whether this is true or not matters little, be-

cause it is thinking so that does the damage. Such a leader views his/her primary task to be transferring a tremendous store of knowledge to the members of the group, who need firm guidance on the road to knowledge. However, this expert may instruct in a kindly way, using the mechanics of democratic group leadership.

At the other extreme is the leader who considers himself/herself as merely a catalyst. This type doesn't know, or claims not to know, much in particular about the subject matter. This leader says that he/she has no ideas as to how to approach the subject. He/she sees his/her mission as merely one of being present, being understanding, keeping the peace, and reflecting the contributions of the group members.

We can question the effectiveness of both extremes. The leader who gets the best results takes a position somewhere between the extremes.

YOUR TASK IS TO PROMOTE LEARNING

As a conference leader, you'll be a teacher in a sense, with your own value judgments. Look at your task as that of helping the group members to learn—from one another, from outside sources, from your own experiences with the subject. Make learning as easy as possible by keeping the environment free of threat and destructive frustration. Try to help the participants understand the subject being discussed and try to facilitate group interaction.

Respect each member of the group as an individual, with rights and needs like your own. Each member wants to learn more about the subject. Your job as conference leader is to create the "climate" that will let each learn to the best of his/her ability. It's important that you understand the group members' interests as well as your own.

Fig. 12-2. As a conference leader, respect each member of the group as an individual, with rights and needs like your own.

AS A CONFERENCE LEADER, DO—

● Study and prepare for the situation carefully. Consider objectives in view of the group make-up.

● In your introduction, relieve initial tensions—both yours and the group's, establish issues clearly, motivate the group to help you, and be reasonably brief.

● Listen carefully—to understand rather than to evaluate.

● In your response to a comment, repeat to make sure you understand the point of the comment; reflect emotional as well as thought content; help the group to understand the point being made; tie the comment to the subject; and relate the comment to other pertinent comments.

● Utilize thought-stimulating questions instead of leading or loaded ones.

● Keep the discussion in the general area of the subject and moving forward.

● Summarize occasionally after a major question or issue has been discussed and again at the end of the session. Or, you may ask one of the group to do this for you.

● Go into enough detail in your concluding remarks that the group members don't leave with the feeling that they spent their time arriving nowhere.

AS A CONFERENCE LEADER, DON'T—

● Lecture the group on the subject or on conference technique.

● Pose so many questions or issues at once that confusion results.

● Call on individuals directly.

◄ Promoting Meetings ►

The relative success of any meeting depends on the number of people who attend. That's why it's so important to use all available communication channels to promote the meeting, thereby letting people know about it and encouraging them to attend.

A meeting should be worth promoting. It's not fair to your audience to build up a meeting in glowing terms and then fail to make it worth the time and trouble to attend.

NEWS STORIES MOST WIDELY USED

News stories are the most widely used promotion

tool. They are most effective when they are well written and released at the proper time.

As a general rule, release the first announcement story about four weeks before the meeting, giving the date, time, place, and a brief paragraph describing the purpose. This story alerts the readers, and it gives them time to arrange their schedules so they can attend.

A second story released three weeks before the meeting can go into a little more detail. You might write a third story the following week covering a few more details or another aspect of the program, but save most of your ammunition until the week before the meeting and then shoot for the moon. A daily newspaper in your area might carry several stories that week.

Sometimes you may have to dig deep for three or four stories about a program that covers only one or two topics. Try to slant each story around a different angle of the meeting. Be sure to include details about date, time, and location in each story.

(See "The Advance Story" and "The Follow-Up Story" in Chapter 3 for more ideas on using the press to promote a meeting.)

PHOTOGRAPHS CATCH ATTENTION

Most editors will appreciate a photograph or two to help promote a meeting. One example of such a photo might be a shot of an automatic feeding setup that farmers can see on a tour. Or, the picture might be a head and shoulders of the principal speaker, or better yet a picture of the speaker doing something that ties in with the speech or a demonstration.

RADIO WORKS WELL

You can use your own radio program for brief, catchy spot announcements about a forthcoming meeting. Even if you don't have your own show, the chances are good that your local stations will make short announcements of upcoming events.

Start radio promotion about two weeks before the meeting, but again build up interest over time and save the details for the week preceding the meeting.

Apply the same rules for preparing radio copy as for writing advance news stories for the press. Interpret the meeting program in terms of audience interest. Point out what persons attending can learn or why they will benefit.

Live or taped interviews can be an effective way to promote meetings on radio. Interview one of the speakers, or farmers if you are promoting a tour. Describe what people will see and learn if they attend.

Fig. 12-3. Use all available communication channels to promote your meetings.

A beeper telephone news report from the meeting site can be effective. Time your call to the station news department so that it can be aired just before the start of the meeting, activity, or event. As you report, try to capture the spirit and excitement of the event. In that way you may motivate the listener to come to the meeting.

HOW TO USE TELEVISION

Television is unsurpassed for some events but has limited value for promoting county meetings. Few counties have their own local TV stations. Metropolitan stations are seldom interested in meetings out in the counties.

A few TV stations, however, produce farm or home programs for the same audience you're trying to reach. The directors of these shows often are willing to use your spot announcements, along with a picture or two if you supply them.

Community antenna or cable television may be better suited for promoting meetings than is broadcast television. Many CATV systems have channels devoted exclusively to local announcements. Take a copy of your program to the local cable service and ask the company to help promote your meeting.

HAVE MASS MEDIA REPRESENTED

Invite representatives of all mass media to your meetings. At first they may give only medium support, but in the long run, they'll get to know better both you and what you are trying to do. Such relationships eventually pay off.

DIRECT MAIL IS EFFECTIVE

Another promotion idea that you'll want to consider involves direct mail post cards or flyers. (For more information on this topic, see Chapter 5.) Get direct mail pieces into the post office at least a week or two before the meeting.

POSTERS AND ANNOUNCEMENTS

Two other promotion possibilities are posters and announcements at other meetings. Even if your talent for drawing limits you to straight lines, you can still make an effective poster. Use bright, cheery colors that will attract attention. Post announcements in your office, the bank, and grocery, hardware, feed, drug, and department stores.

Announcing the forthcoming program or event at other meetings gives interested persons a chance to ask questions about the program.

COVERING THE MEETING

Even when the meeting is over and chalked up as a success, you'll have another job to do—follow-up coverage. This coverage is designed for the person who couldn't attend, but who is interested to know if the speaker's message was relevant to his/her own problems.

If you've asked the media editors or reporters to attend and they do, your coverage problem is solved. But if they don't come, you'll have to sit down and hammer out a story yourself.

Report highlights of the meeting on your radio program, or give copies of your news stories to the station manager or news editor. Also send copies to the farm or news director of your local TV station.

When a meeting or tour offers good picture possibilities, shoot some photos for the local papers or TV station.

HOLDING A PRESS CONFERENCE

Another way to let people know about your meeting is to have a press conference, using a speaker or leader from the meeting.

A cardinal rule, however, is to be absolutely sure you have news to offer that will justify asking reporters to attend. Limit the subject to a significant topic. If you're unsure about media reaction, get in touch with an editor or a broadcaster and ask him/her if he/she would be interested.

Assuming you do have media interest, think well in advance about whom to invite. Prepare a list of newspaper, radio, and television people—their names, affiliations, telephone numbers, and addresses. Keep the list, too, for it may come in handy for later events.

In large cities, you probably can get the wire services to carry a note about your press conference on their daily calendars a day or two before. It's also a good idea to contact the people on your list by telephone and maybe even in person. If you make a personal visit, don't just drop in unannounced. Call first. Be ready to say who can provide a local angle if there's likely to be such a need and in case the media people don't want to talk to the individuals you propose to run the conference.

Ask the news people how you can best help them record the press conference as to items such as title cards, staging, lights, lecterns on which to attach microphones, repetition of questions, and so on.

Be sure to schedule the press conference for a time when your experts don't have another assignment at the meeting. Make signs to help media people find the press conference—at a place near your meeting's press room, but not in it in order to minimize paper shuffling, telephone calls, and other disruptions during the conference itself.

Plan to hold the conference between 10:00 a.m. and 2:00 p.m., for best coverage by evening news outlets. Also remember that Fridays are usually "bad"; Mondays are usually "good." That's because news quantity tends to collect toward the end of a week and tapers off over weekends, so there is less competition for news space at the beginning of the week.

Have biographical sketches available about meeting speakers and especially about participants at the press conference. But limit the number of principals at the press conference to one person if you can, although that person may want some experts to help field certain questions. Also, limit formal remarks to a brief introductory statement, then let the media people ask questions. Remember that media people are busy and need to get the angle they want quickly.

As you plan whether or not to hold a press conference, consider that you might augment or substitute one with radio actualities. If you do that, develop a schedule with speakers so you can go to a quiet place with them and tape brief summaries of their main points rather than recording them during their presentations. Even if you go ahead with a press conference, you'll have obtained useful material that many stations might like to have. It also makes speakers who won't be participants in the press conference feel better about being left out.

When you hold the press conference, limit the audience to news media and trade writers, such as newsletter editors or people who are going to publish or broadcast the news to many, many others. Avoid a roomful of onlookers—the curious and interlopers with self-interests. Otherwise, they might dominate the conference by asking all the questions and then never send a word to others.

Open the conference by making it clear that questions should come only from the press. You might even want questioners to identify themselves and their news organizations. Tell the media who the spokesperson will be, note what materials you have available (or pass them out early), and let the conference take its course.

Among the handouts you should have available are the statement or press release announced by the spokesperson, any useful major speeches from the meeting itself—full texts, if possible—and a sheet with names, titles, addresses, and affiliations of your conference experts.

Much of the time, the senior news representative present will note when colleagues have run out of questions and will close the conference with a "thank you." You may have to remind him/her to do that. Otherwise, watch for a lag in questions yourself and slip a note to your spokesperson who can thank the media for coming, thereby closing the conference.

You can evaluate the success of the press conference by keeping watch for media reports. Also, be sure to make mental notes at least of things you would do differently next time so you can make adjustments when you have a press conference again.

◀ Staging Your Presentation ▶

Next time you watch TV, take the trouble to read the credit lines. You'll find the names of directors listed in the credits. Directors are proud of their role in TV productions. Can you be as proud of your meeting presentation?

If you've ever watched directors of TV shows, you may have seen them simulate the picture by forming a frame with their hands. Have you ever taken a similar look at your presentation from your audience's point of view? Many things affect their viewpoint, but one of the most important is the condition of the meeting room.

HAVE A CLEAN AND COMFORTABLE MEETING ROOM

You may not be able to hold your meeting in an ideal room. Locations may vary from a church basement, a neighborhood community hall, a classroom, or a courtroom to a modern auditorium. But in any case, you can keep the presentation area free from annoying clutter.

Clear off messy bulletin boards. Take down sagging or discolored decorations and remove piles of odd items from the corners. Focus attention on the program as the center of attraction. Don't let anything detract from it, such as a bright glare from front windows behind the podium or unused projection screens. Be sure to pull shades or drapes over glaring windows before the meeting begins.

Arrange the chairs to give listeners plenty of space between rows so that they can relax and enjoy the program. Provide enough aisles so that latecomers can get to vacant seats without having to work their way over numerous pairs of feet. Better yet, have an usher to help locate empty seats and keep late arrival confusion at a minimum.

Above all, be sure the room has good ventilation. Even though the room temperature may be normal when the room is empty, don't forget that it may become uncomfortably warm and stuffy after the audience arrives. It's better to let the early arrivals be somewhat cool than to lose your audience halfway through the program because the room has become too hot. An usher or assistant can check the room temperature after the meeting is underway and increase the heat or open the windows as necessary.

If fans or ventilation systems are too noisy, they may drown out speakers, so make sure in advance whether you'll need a public address system. If so, put it in place before the meeting begins.

Be sure there are plenty of places to hang coats. People have enough trouble following a presentation without having to wrestle with heavy coats.

CHECK THE PRESENTATION AREA

Look at the presentation area as your audience will see it. If there is a platform or podium, is it large enough? Will it be noisy? A platform that is too small or noisy is worse than none at all. You can use strips of carpet or other material to stifle the noise.

Before the meeting begins, make sure the chalkboards are clean and that chalk and erasers are handy. Use the thick jumbo chalk sticks; they make heavy lines that can be seen from the back row. Similarly, check over any portable equipment that may be used.

Easels and stands should be sturdy and shouldn't slide or collapse under a load. Seeing an easel col-

Fig. 12-4. Before a meeting begins, check to make sure all audio-visuals and supplementary materials are on hand and in a usable condition.

lapse may be amusing, but it will ruin your rapport with the audience. Most of what you have presented will be lost, and it will take some time to recapture the group's interest.

Flannelboards should be clean and free from lint. You can raise the flannel nap by brushing with a stiff brush.

Don't forget lighting. Chalkboards, flannelboards, and charts need to be well lighted so that they can be easily seen. Investment in several clamp-on lamps with goose-neck stems may be well worthwhile. If you use such lamps, be sure the light shines on the material and not into the eyes of the audience.

Above all, use large, legible print or writing. A ½-inch letter is visible for only 16 feet. A 1-inch letter can be seen from 32 feet. A 2-inch letter can be seen from 64 feet. Practice using thick lines. This will pay dividends.

Last, but most important, check your projection equipment. Get a solid, heavy table or stand to support the equipment. Allow enough space around a projector located in an aisle so that people won't bump into it. Use an adequate length of heavy-duty extension cord. Tape it to the floor so people won't trip over it. If you're in a strange room, have someone check the fuses to be sure they can carry the load. Determine the screen location before the meeting starts. If it's not possible to leave the screen up, lower it and then raise it when needed.

If the whole screen needs moving, locate the position of the feet on the floor with bits of masking tape. With these guide marks you can relocate your screen quickly and accurately. Then align your projector and focus it. Thread motion picture film and run a few frames to check your work. Be sure to have a spare projection lamp beside the projector. Check the sound level.

LAST-MINUTE DETAILS

A few last-minute details are still in order. Run through your presentation with an assistant and other participants before the scheduled starting time and before the audience arrives. Avoid using audio cues to turn on lights, change slides, or perform other chores. Arrange silent signals or gestures, or use a small pilot light for an indicator.

Remember: Start on time, keep on time, and end on time.

◀ In Case of Emergency ▶

You are all set to go, and the main speaker can't get there. What can you do? First, don't panic. Second, use the resources available in the group. There may be individuals present who can talk on particular points that are pertinent. You may form a sharing panel, with several people discussing their experiences or thoughts on the topic of the day. Or, you can always divide the audience into buzz groups for 10 minutes to formulate questions they would like to have discussed. After the 10 minutes, lead a discussion of these points.

Here are some other suggestions:

1. If the speaker can't get there (for reasons other than health), try to set up a tele-lecture system, whereby the speaker can talk to the group via the telephone from wherever he/she might be (including a public phone booth at an airport). If you're not acquainted with this system, talk with your local telephone officials about it.

2. Talk to the stranded speaker by phone. Ask for

the key points he/she was planning to make. Your secretary should also be on the line to take down the main points in shorthand and transcribe them. Read them at the meeting.

3. Use films and slides. Another possibility would be to use music to entertain during a delay in starting time or to fill in during mechanical breakdowns.

4. In case of an emergency, inform your program committee members or assistants of the emergency procedures. This will put you in a much better position to handle the situation.

◀ Selected Meeting Techniques ▶

Variety in program presentation heightens audience interest and promotes audience participation. There are nine meeting techniques that have been used with good results by many groups and organizations. These are: lectures, symposiums, panels, forums, buzz sessions, skits, role playing, discussion groups, and question-and-answer periods.

Lectures. A lecture consists of a speaker making a presentation from the front of the room or on a platform, using primarily an oral presentation which may be supplemented by visuals such as slides, chalkboard, magnet board, flipchart, flannelboard, etc.

The lecture method is most useful when the objective is to present just one point of view, when time is limited, or when the primary objective is to get the views of an "expert." It's effective in presenting new material, relatively easy to organize, and suitable for a large audience.

Symposiums. A series of prepared speeches by several speakers, usually considered experts in areas related to the general topic under discussion, constitutes a symposium. Some visuals (those mentioned under "Lectures") may be used, but presentations are usually oral only. A moderator presides.

The symposium is most useful when the objective is to present several aspects of a subject; it's particularly good when the objective is to present several viewpoints, air several sides of a controversial issue, or clarify several facets of a complex problem.

Panels. A panel is made up of several people seated around a table who carry on a dialogue among themselves. One member of the group serves as the moderator.

A panel is especially good when the objective is to develop a topic from several viewpoints. A panel gives the audience an understanding of the various parts of a problem, and it helps them to identify or explore the problem.

Forums. A formal presentation by one or more

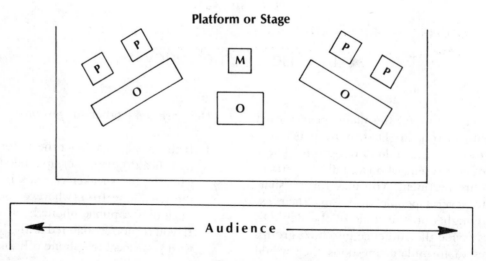

Fig. 12-5. A suggested physical arrangement for a panel: M—moderator, P—panel members, O—microphone.

speakers followed by questions and comments from the audience is a forum.

A forum permits audience participation and discussion in a large meeting. It might get some thoughts troubling the audience out into the open. It also encourages the development of a topic from several viewpoints. It can help jell action or form community opinion.

Buzz sessions. A buzz session consists of small groups (usually four to seven members per group), which are made up of members from the audience. Buzz groups may be organized formally with a chairperson and recorder, or they may operate without a leader. They may be given a specific topic or problem to discuss, or they may be used to develop questions, indicate areas of interest, or offer evaluations. A buzz session often follows a formal presentation.

A buzz session involves every member of an audience, regardless of size, and it permits widespread participation and identification in a large discussion. It promotes individual identification with the larger whole.

Skits. A skit is a short, rehearsed dramatic presentation involving two or more persons, who usually use a prepared script. It makes up only one small part of a total program, coming at any point in the program before the discussion period.

The skit is used to dramatize an idea, introduce a topic for discussion, highlight a situation, create audience interest, or insert an entertainment spot in the program.

Role playing. Role playing differs from a skit in that the "actors" make up their parts as they go along.

There's no script prepared in advance. Role playing is usually followed by a discussion from the audience or by other members of the group.

Role playing is often useful in the area of human relations where the problems can be particularly emotional and where there may be a desire to change attitudes. It's been particularly effective in getting people to see the other side of a question, and it's been used extensively in mental health therapy. It's most effective with small groups.

Discussion groups. A discussion group is made up of a small group of individuals (not more than 20 members), who sit around a table or in a semicircle, with one person serving as discussion leader. There is usually no formal presentation—just the leader identifying the problem or setting the stage. Informality is important. The leader doesn't usually designate who will speak or "have the floor"; however, he/she should guide the discussion and maintain order.

Discussion groups are able to pool the thoughts of all the participants and hopefully arrive at group decisions. The combination of lecture-discussion is the preferred meeting technique of most adults.

Question-and-answer periods. In a question-and-answer period, members of the audience submit oral or written questions to the speaker. An announcement explaining that questions will be considered and in what form is usually made before the speech.

A question-and-answer period is a good way to get answers to questions not answered by the speaker. Questions can introduce some informality and get audience participation in an otherwise strictly formal and one-way program.

Photo / Art Credits

Bill Ballard, North Carolina State University: Figs. 12-1, 12-2, 12-3, 12-4.
William L. Carpenter, North Carolina State University: Fig. 12-5.

13 NEW COMMUNICATION TECHNOLOGY

◄ Changing Channels ►

We're entering the twenty-first century riding a tide of rapidly changing communication opportunities.

Through the years our *sources* may have remained somewhat unchanged, our *messages* have always been undergoing evolutionary developments, our *channels* have been undergoing modest adjustments, and our *receivers* have steadily become more sophisticated. (See the SMCR model described in Chapter 1.)

But look out! In the 1980s the *channels* are undergoing such rapid changes that you may wake up some morning ill-equipped to communicate with your clientele. This chapter attempts to deal with some of those changing channels. The following is a list of some of these new channels. Each is described briefly, along with a new words section, later in this chapter.

- ● Low power television service (LPTS)
- ● Cable television
- ● Telephone and video conferencing
- ● Dial access (Teletip)
- ● Computer networks
- ● Green thumb boxes (viewdata or teletext)
- ● Microcomputers
- ● Electronic mail
- ● Satellite transmission and receiving

Aha, you say, those aren't new. All that technology has been bumping around for several years! Yes, but we're just beginning to put it all together. We're just beginning to reap the benefits of these massive leaps forward. By using modern technology, we can now improve information transfer and communication programs.

◄ Instant Information ►

"Instant information" 20 years ago was a dream. We speculated that if we could provide extension agents and mass media with capsules of "instant information" when they wanted it—at their convenience—we could greatly improve our communications efforts. The transmission, storage, and delivery tools for "instant information" weren't available 20 years ago; but they're available now!

Prepared by **Eldon E. Fredericks,** head, Department of Agricultural Information and Department of Audio-Visual Production, Purdue University.

An estimated 400,000[1] microcomputers were owned by individuals in mid-1981. Predictions that more than 400,000 word processors will be in offices by 1990 are common; and Haldeman[2] projects that 3 million residences will have some type of home terminal services by 1985, with growth to 12 million by 1990 and 25 million by 1995. For a relatively small fee, a microcomputer can be programmed to call a local phone number, receive stock market quotations, analyse the performance of specific stocks, and recommend a buy-or-sell strategy.

Grain and livestock market information is available to microcomputer subscribers from a news service called Instant Update, located in Cedar Rapids, Iowa. The Michigan Farm Radio Network in Milan, Michigan, will soon be providing market information via computer and telephone link to better serve its farmer audience. The computer in a home learning center will deliver and receive many kinds of instant information, from the latest football scores to the effects of weather conditions on world soybean prices.

[1]"Apple, The Personal Computer Magazine and Catalog," Vol. 2, No. 1, 1981.

[2]"Home Terminal Systems," a prospectus for a strategic master plan, Lloyd H. Haldeman, president, HVC Corporation, Dallas, Texas, May, 1981.

◀ Tools to Reach Masses with ▶ Specific Information

How can our education and information programs provide that precise "instant information" to the many clients demanding specific knowledge?

Traditional educational delivery methods will be questioned as we enter the twenty-first century. These include:

- Meetings
- General public radio and television programs
- Press releases to mass media
- Field days

These stalwart methods, while still important, will be supplemented with new communication technology by those who provide technical information to a changing clientele.

NEW TECHNOLOGY IN USE

We don't have to gaze very far into the future to see big changes in education and information programs. Many land-grant universities and their extension services and experiment stations are using computers for a variety of communication and decision-making tasks. For example:

- Michigan State University TELPLAN and COMNET
- Purdue University FACTS
- University of Nebraska AGNET
- Virginia Polytechnic Institute CMN
- Pest management programs at various universities

These examples represent just a few of these programs. There are many others, and there will be new ones added regularly.

Agricultural information offices at several universities and USDA agencies have adapted computer programs to transmit news and feature releases to mass media and to county extension offices.[3] Various methods are being developed at Michigan State University, Oregon State University, Purdue University, the University of Nebraska, and the USDA North Central Region Agricultural Research Information Office in Peoria, Illinois. Others are investigating the opportunities and considering the potential benefits. Conventional reproduction and distribution costs compared with the speed and newness of the computer offer incentives for change.

In almost all cases, computer systems developed because there was a need to improve the knowledge bases to help someone make sound decisions. Educational programs have ranged from balancing livestock rations to determining the cost/benefit ratio of chemical applications to control an insect or disease infestation. Other programs have been introduced to deal with inflation and energy consumption in the home. Now, communication via computer adds another dimension to our information transfer systems.

Small, relatively inexpensive computer terminals

[3]Eldon E. Fredericks, Stephen B. Harsh, Edward Rister, and Deb Vergeson, "News Releases Delivered by Computer," *ACE Quarterly Journal,* June, 1979.

have replaced the cumbersome models of a few years ago. Knowledge workers today can carry terminals in their cars or on long distance trips to keep in touch by phone with their routine activities. Electronic mail is widely available through several commercial networks.

Airlines, banks, manufacturers, and other businesses have automated their information transfer systems. The complex flight reservations, fund transfers, and production scheduling represent but three giant steps in the development of instant information.

Fig. 13-2. County extension offices in many states have computer terminals or microcomputers hooked through networks to their land-grant universities. The Fast Agricultural Communications Terminal System (FACTS) in Indiana links the 92 county offices and 10 area administrative offices to Purdue University. The stand-alone units in each office provide management information for clients through computer programs developed by Purdue specialists. They also keep records, compile mailing lists, and serve as word processing systems.

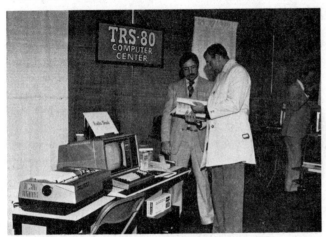

Fig. 13-1. Communicators are beginning to update their skills in a new way to transmit information across the country in a matter of seconds rather than days. Microcomputers linked with main frame computers and telephone lines allow sending and receiving complete documents as well as one-page press releases. As more homeowners purchase microcomputers, direct delivery to consumers will be easier for the computer-trained communicators.

KNOWLEDGE BUSINESS

According to a Booze Allen office automation study conducted in 1979, 55 percent of the U.S. workers are classified as information or "knowledge" workers. Alvin Toffler[4] in *The Third Wave* says, "The average U.S. factory worker is supported by a $25,000 investment in equipment while the U.S. office worker has an average of $500 to $1,000 worth of old typewriters and adding machines." Toffler

also cites the decrease in numbers of factory workers and the increase in numbers of office workers. He says, "The need for information has mushroomed so wildly that no army of clerks, typists, and secretaries, no matter how large or hard-working, can possibly cope with it."

The knowledge explosion we heard so much about in the 1970s may have been a "reproduction explosion." We learned to publish more information in a variety of media for specific audiences. In the 1980s we're going through an "electronic explosion." Only with the help of computers will we be able to file, sort, retrieve, and use the precise "knowledge" to fit the specific question.

The computer may provide some hope for dealing with the vast array of information flooding our desks every day. Computers can help us file and find the precise information needed to answer very specific questions from our clientele. Vast data bases can be searched by computer terminal and telephone lines. Such a search can be completed in your home or office rather than in a major library. When it is completed, you can receive a printed copy of the results of your search.

[4]Alvin Toffler, *The Third Wave,* Bantom, 1981.

◀ The Knowledge Office of Today ▶

The powerful communication application of computer devices is just becoming apparent to the masses. Word processors (sometimes called power typewriters) will soon become the standard for many offices. Within two to five years a more complete information flow will likely be handled electronically.

With networking, the word processor will serve as a dictation and typing station, duplication center, and a filing and distribution system.

A shared logic word processing system has been installed at Michigan State University's College of Agriculture and Natural Resources. This cooperative venture between the college, the Agricultural Experiment Station, and the Cooperative Extension Service will eventually move documents from office to office with limited paper copies. When first proposed in 1979, the concept called for a system capable of moving a publication from author to editor to county extension office, electronically. Words don't touch paper until they reach their final destination—unless someone requests a paper copy.

The Michigan State system, when fully operational, will work like this: An extension specialist in the Crops and Soils Department will prepare the annual *Weed Control Manual* on the shared logic typing terminal. That manuscript will move electronically to a publications editor. After editing, the editor will return it to the author, electronically, for approval. Then back to the editor, who can release the manuscript to the communication computer for automatic transmission to each county extension office and the 15 branch experiment stations. The editor also will send the manuscript, electronically, to the typesetter or printer for preparation of the final publication.

With this distribution system, fewer copies will be printed and stored. And, if a major change is needed, it can be entered easily on the word processing equipment and transmitted immediately to anyone on the system. Revising this publication the following year will be a breeze because the entire manuscript will be stored on the system.

STAFF INVOLVEMENT

The drastic changes in office procedures described here may create problems for knowledge workers and their support staff. Gail McClure, head of Communications Resources at the University of Minnesota Institute of Agriculture, Forestry, and Home Economics, is concerned about preparing people to deal with technological innovations. McClure says, "People and their reaction and support of change must be a major part of every new project or activity contemplated."

Readers of this handbook may wish to become familiar with the opportunities and the problems associated with changes in the way people are receiving information. Change has always been a part of our lives, and we learn to cope; however, those changes came less rapidly in the past. Educators, and specifically extension workers, have been called "change

agents." Are we willing to change the way we operate as eagerly as we recommend change to our clientele? Computers and other communications technology have already "invaded" our communities. We can help that invasion turn into a successful beachhead and a profitable occupation. Or, we can throw rocks and hope the new methods will go away.

LEARN WHERE YOU CAN

Be aware of opportunities in your community. Learn as much from the people using computers as you can. From them you can seek training in other places.

Find out how the local bank uses computers to transfer funds. Talk to the manager and workers to see how their jobs have been changed and how these changes could have been made easier.

Ask airline ticket agents how much training they received when the shift was made to computer flight scheduling. Ask them to compare the time they would spend working out a trip for you on the computer versus going through the rate and schedule books of yesterday.

Talk with the manager of the phone company about the directory assistance program. Find out how the local and long distance operators can get phone numbers for you by querying a computer. Ask them to compare that to a hand search through the directory.

As providers of information, we need to stay in touch with current methods. Read and observe all you can about emerging communication technology. Forrest D. Cress and John D. Fox,[5] communication specialists at the University of California, described some of these changes in a paper prepared for their extension service in September, 1980. Similar studies have been conducted at the University of Minnesota[6] and at Cornell University.[7]

Some of you may want to look into computer training. Most of you probably don't have the desire or the need to become computer scientists or even programmers. However, a course in computer literacy at the local high school, community college, or other learning center might be worthwhile. You

[5]John D. Fox and Forrest D. Cress, "New Technologies: Implications for University of California Cooperative Extension," September, 1980.

[6]Sam Swan, Linda Camp, and Neil Anderson, "Communication Technology Report," Agricultural Extension Service, University of Minnesota, 1980.

[7]"Implementing a Word Processing Communication System for the College of Agriculture and Life Sciences, Experiment Stations, and Cooperative Extension Locations," report prepared by Word Processing Committee, Cornell University, David Dik, Chairman, February, 1981.

could also visit microcomputer sales stores and talk with the young people always gathered around those machines.

Perhaps the next edition of this handbook will be available on computer diskette or as a videodisk with visual support programs. You may have the opportunity to work with a terminal keyboard to review the text of the fifth edition.

We're just beginning an exciting era. Happy communicating!

◄ Some Examples of ►
New Communication Technology

CABLE TELEVISION

Early cable installations were designed to bring a reliable signal to remote and isolated areas of the country. They carried the major network channels and perhaps a local public station. Today, cable television, with the help of satellites, microwave, and other signal carriers, brings as many as 50 or more channels into remote and highly populated areas. While nearly all systems are strung together by wire or cable, we are beginning to see individual homes and businesses installing small satellite receivers. "Cableless" television may soon come to rural areas. And, in several cities a two-way cable installation is already available. CUBE in Columbus, Ohio, permits viewers to talk back and respond to questions.

LOW POWER TELEVISION
SERVICE (LPTS)

On September 9, 1980, the FCC adopted a "Notice of Proposed Rulemaking," authorizing a new class of broadcast stations called Low Power Television Service, which would be allowed to originate an unlimited amount of programming. LPTS will operate on a secondary, non-interfering basis to regular full service TV broadcast operations. The range of LPTS will be approximately 10 to 12 miles in diameter. It will include existing TV translators. LPTS will also be able to encode program material from any source, so long as program rights are first obtained.

LPTS is the first new broadcast service considered by the FCC in more than 20 years. For small communities, both rural and urban, it means an opportunity to have local TV. The costs of LPTS are much lower than full service TV. The opportunities are almost limitless: local news, public meetings, community events, educational programs, and school board meetings could keep people informed about local issues and decisions.

TELEPHONE AND VIDEO
CONFERENCING

Holding meetings in several locations with speakers or groups connected by telephone lines has become fairly common in recent years. The energy shortages and resultant higher costs of travel hastened this use of electronic conferencing. The University of Wisconsin is recognized as a leader in the development of this teaching technique.

Video conferencing, the transmission of a motion or active picture (usually in one direction), is becoming economical with the advent of satellites. Usually a two-way telephone hookup permits audio conversation between all sites in such a conference, with the video or picture eminating from one location. More information may be obtained from these sources:

Center for Interactive Programs
Old Radio Hall
975 Observatory Drive
Madison, Wisconsin 53706
Phone: 608-262-4342

Bell and Howell Satellite Network
985 L'Enfant Plaza, SW, North Building
Washington, D.C. 20024
Phone: 202-484-9270

Appalachian Community Service Network
1200 New Hampshire Avenue, NW, Suite 240
Washington, D.C. 20036
Phone: 202-331-8100

Public Service Satellite Consortium
1660 L Street, NW, Suite 907
Washington, D.C. 20036
Phone: 202-331-1154

DIAL ACCESS (TELETIP)

Recorded messages can often answer questions posed by phone callers. Many extension offices have

Fig. 13-3. Recorded messages accessed by telephone have become a popular educational technique in many subject areas. This statewide system, operated by the North Carolina Cooperative Extension Service, contains more than 1,000 messages. Usage has ranged as high as 1,000 calls in a single day.

these devices, which provide answers to redundant questions, do it faster, and relieve the staff of this repetitive chore. Hospitals, tourist information centers, and many other public service agencies have taken advantage of this modern technology to provide information to an information-seeking population.

The extension service at North Carolina State University applied this technology early. In 3½ years, more than one-half million calls have been received by the 1000-message system.

COMPUTER NETWORKS

When first developed, computers generally were too large and too expensive for one owner. Thus, time-sharing was implemented as a way to spread the use and the cost of these monsters. As they became smaller and more affordable, more companies and agencies found they could justify their own computers. However, the ability to "communicate" with another computer in another branch or division or in another company was still highly desirable. Such networking has grown into a service offered by several national commercial companies. Some examples of these include COMPUSERVE in Columbus, Ohio; SOURCE in McLean, Virginia; TYMNET in Cupertino, California; TELENET in Vienna, Virginia; DIALCOM in Silver Springs, Maryland; and a host of others.

GREEN THUMB BOX

Other countries, notably France, Great Britain, and Canada, have developed electronic transmission of information for homes more rapidly than the United States. Such information includes newspaper stories, classified advertisements, yellow pages, phone directories, and some postal services.

The University of Kentucky Extension Service, with assistance from the U.S. Department of Agriculture and the U.S. Department of Commerce, pilot tested such a concept called "Green Thumb" in 1980–81. Approximately 200 farm families in two Kentucky counties were given special "green thumb boxes" that permitted access to a small county-based computer. By networking, information from a large computer at the university provides information from extension specialists, the Chicago Board of Trade, the National Weather Service, and other trade associations. The initial trial period ended in mid-1981, with several evaluations of the system anticipated by 1982.

MICROCOMPUTERS

The electronic revolution has led to compressing millions of transistors into something that resembles a portable typewriter. When hooked to an ordinary television set, the microcomputer, sometimes called a personal or home computer, comes alive and plays games, displays graphics, prepares complicated decision-making procedures, and communicates with data banks through various networks.

The microcomputer may be the simple green thumb box variety, or it may have its own storage or memory to give its operator the ability to make mathematical calculations or to manipulate letters into words in the form of a word processor.

Fig. 13-4. Many companies manufacture and sell microcomputers for homes, offices, and small businesses. As farmers are purchasing these units, a market for software programs is growing very quickly. Record keeping which is specific to the needs of dairy farming, swine production, crop production, weather and pest management, and a host of other potential data base manipulations is gaining momentum.

The interactive capability of microcomputers with cable, video disks and cassettes, teletext or viewdata terminals, dial access systems, and teleconferencing systems is combining technologies for better efficiency and more applications.

ELECTRONIC MAIL

Using computers and networks, we can move messages without paper from one location to another. These messages can be transmitted back and forth at the convenience of the receivers. But, unlike the postal service or other document carriers, the electronic message can be available in another part of the world within a millisecond of release. And, it can be distributed to as many persons as have receivers with the same coding identifiers.

SATELLITE TRANSMISSION AND RECEIVING

Satellites help make it all possible. We no longer need to string miles of cable around the world like a kitten with a ball of yarn. Now we merely beam a signal up to a satellite and that same signal rebounds to a receiver in another location. Pictures, voices, and digits all travel with the ease of lightning through the sky. A flyer from the PSSC describes satellite capabilities as follows: "Communications satellites provide instant and reliable contact between any two or more points on earth, far beyond the capabilities of terrestrial methods."

The basic characteristics of satellite systems are:

1. They consist of an earth station at both ends and a satellite in the middle.

2. They're flexible. Points can be added or subtracted anywhere in the network by adding or removing an earth station.

3. They're distance-insensitive. Any two points are equidistant to the satellite. This brings down the cost of communicating between two distant points, from one point to hundreds or thousands, or from many points to one or more points.

4. They offer increased choices to public service organizations in the areas that can be made available—including video, audio, and facsimile and data transmission services.

Larger, more efficient satellites scheduled for launch in the 1980s will result in smaller, more economical earth stations. This will provide improved communications with flexibility and lower cost.

◀ A Few Selected Terms ▶

Acoustic coupler. A device to connect a computer or a terminal to a telephone handset. Allows sending and receiving data as tones. (See MODEM.)

ASCII. (pronounced ass-key-2) An abbreviation for American Standard Code for Information Interchange. The standard which permits data transmission between computers.

Baud rate. The rate of transmission of signals, usually in bits per second. Normally speeds of 110, 300, 1,200 or 2,400 are used, with limitation being the ability of the phone line. When equipment is wired without phone lines, the rate may be 9,600 or more. (300 baud is approximately 30 characters per second.)

Bit. A unit of information, either 0 or 1. (Comes in binary digits.)

Byte. A group of binary digits usually equated with a character. One byte roughly equals one character.

Central processing unit (CPU). The computer excluding any peripheral devices.

Character. A letter, numeral, or symbol used to represent information.

Command. An order given to a computer system, usually by a user through a keyboard.

Compatibility. The ability of an instruction or language to be understood and used on more than one computer.

Control character. An invisible character inserted to tell a receiving station to perform some function. (Might be paragraph indent, end of page, centering command, etc.)

CRT. A cathode ray tube (looks like a television set) which is used to display information. May be called a VDT (video display terminal). This is one type of peripheral device.

Cursor. A position indicator frequently used on a CRT to indicate where characters will be typed.

Data. The basic elements of information which can be processed or produced by a computer; includes facts, numbers, letters, and symbols.

Dump. To copy the contents of all or part of memory onto some other storage medium.

File. A unit of some records or text. A file might be a news story, a manuscript for publication, or a business letter.

Filename. A designation that identifies a particular file. The user assigns file names that are easy to remember.

Hardware. The physical equipment associated with a computer or other electronic device.

Intelligent terminal. A terminal with some amount of programmable capability so that it can perform certain tasks without being connected to a computer.

Interactive. A system that performs processing or problem-solving tasks by "talking" with the user.

Interface. A connection, such as the interface between a modem and a terminal.

K. An abbreviation for the prefix "kilo." In common usage it means about "1,000." A 48K memory is capable of holding approximately 48,000 bytes (characters) of information.

Load. To place data or programs into a computer's internal storage for processing or manipulation.

Memory. (1) The alterable storage space in a computer. (2) A device in which data can be stored and from which they can be retrieved. Might be floppy disk, hard disk, or magnetic tape.

MODEM. A modulation/demodulation device that enables computers and terminals to communicate over telephone lines. (It is similar to an acoustical coupler but doesn't use the handset and gives more dependable connection.)

Network. The interconnection of several communication devices, terminals, and/or computers. Networking will become the wave of the future in communication, word processing, and office automation!

Peripheral equipment. Any unit distinct from the central processing unit (CPU), which provides outside storage or communication. (May be a printer, disk storage unit, CRT, etc.)

Remote. Located away from a local computer, terminal, or other device.

RS-232. A technical specification published by the Electronic Industries Association to establish the interface requirements between MODEMS and terminals or computers. (For use in telephone transmission.)

Software. Programs that provide routine instructions to a computer.

Terminal. A peripheral device enabling a user to send data to or receive data from a computer. It may be a CRT, or it may be a paper printer.

User. The programmer or operator of a computer.

Photo / Art Credits

Audio-Visual Production Department, Purdue University: Figs. 13-1, 13-2, 13-4.

Vellie Matthews, North Carolina State University: Fig. 13-3.

14   A MULTI-MEDIA APPROACH

A multi-media approach in extension teaching and dissemination of knowledge can be interpreted in many ways. To some it is creating in paints, yarns, clay, etc. To some it means only multi-screen projected images. To others it's the use of radio, newspapers, television, etc., with little thought about relationships.

But a multi-media pre-plan offers a more effective and efficient approach. It can be a guide to a coordinated interdisciplinary approach which uses a variety of communication media.

Coordinating local programs of study for a multi-media approach is time consuming and long range, hopefully involving many people and organizations.

Ideally, a project or a topic is determined a year in advance. This is usually based on a survey, and/or lay leader analysis to determine need and priority. Once a focus has been established, a rather detailed plan, both in depth and breadth, can be laid out. Determining responsibility for each facet (sometimes as much as a year ahead) is a vital step.

A local leader, a local or area newspaper representative, a radio personality or program director, and a television station manager or news staff person are examples of those who can share ideas with the extension staff to get an effective multi-media approach.

Workshops, leader training meetings, posters, flyers, demonstrations, testimonial support, all can have a part in a successful educational or action project.

Just as skill in using each medium is necessary, so is an overview of the interrelation of one medium to another. Consideration of relative values and effectiveness of each medium with its specific audience is also important in the developmental process.

Newspapers, radio, television, slide tapes, photography, and other media are all part of the educational technology awaiting the uniqueness and creativity of human resources.

Media techniques are a means to an end, not an end in themselves. Hopefully, the end results justify the means. For example, in a small southwest Missouri county, a coordinated mass media series utilizing television, radio, newspaper, direct mail, and a supportive group was used. A diabetic and blood pressure screening clinic was sponsored by county extension clubs, with a week-long clinic in six testing centers over the county.

Pre-alerting methods included:

1. County extension clubs used the subject of diabetes as a study topic.

2. Several clubs provided workers to help conduct the clinics.

3. Local extension staff members prepared radio tapes and sent them to local radio stations.

4. One county extension member developed a special television program for one of the local television stations. Other staff members put together spot announcements for both radio and television.

5. Each club sent out mailed announcements to a random sample throughout the county.

Informal evaluation indicated much of the success of the program was the advance planning and coordinated interdisciplinary approach to effective media use.

Prepared by *Orrine Z. Gregory,* extension information specialist, University of Missouri.

Tangible, measurable results showed a total of 887 persons were screened. Of this number, 95 persons were referred to their family physicians for further evaluation concerning diabetes, and 30, concerning blood pressure problems.

Twenty-five percent of the club membership saw the special television program or spot announcements. Twenty percent heard the radio announcements. Eighty-five percent read the announcements and article in the newspaper. Seventy-five percent of the membership remembered receiving direct mail on the subject. Sixty-five percent of the membership attended the club meeting on the subject. Seventy percent of the membership went to the screening clinic.

Evaluation of this activity confirmed findings based on a similarly coordinated multi-media approach conducted and researched 10 years ago.[1]

Another example was an agronomy specialist in northwest Missouri who recently won top awards in almost all agricultural communications categories and was especially recognized for his portfolio of a multi-media entry.[2]

In southcentral Missouri, a youth program leader maximizes the use of multi-media. He uses radio and television regularly. He prepares news articles for local publication, as well as contributing to a popular regional trade magazine and publishing a tabloid newspaper for extension clientele.[3]

It's the plan! It's the plan that gets the long-time total results. It's the plan that saves time because facts and support materials are coordinated before a project or a program is started. It's the plan that helps a good county or area program to run smoothly through a coordinated multi-media approach.

------

[1]Ellen Jones, "A Ten-Year Follow-up Study of Three Types of Mass Media Methods and Their Effectiveness in Southwest Missouri," April, 1978.

[2]Donald Null, northwest Missouri agronomy specialist at Grant City, received first place awards in radio, newsletters, direct mail, and portfolio categories, University of Missouri Extension Association 1981 Communications Awards.

[3]Youth specialists, Missouri Lakes County extension area.